T0358374

Foundations for Fintech

Global Fintech Institute - World Scientific Series on Fintech

Print ISSN: 2737-5897
Online ISSN: 2737-5900

Series Editors: David LEE Kuo Chuen *(Global Fintech Institute, Singapore &*
Singapore University of Social Sciences, Singapore)
Joseph LIM *(Singapore University of Social Sciences, Singapore)*
PHOON Kok Fai *(Singapore University of Social Sciences, Singapore)*
WANG Yu *(Singapore University of Social Sciences, Singapore)*

In the digital era, emerging technologies such as artificial intelligence, big data, and blockchain have revolutionized people's daily lives and brought many opportunities and challenges to industries. With the increasing demand for talents in the fintech realm, this book series serves as a good guide for practitioners who are seeking to understand the basics of fintech and the applications of different technologies. This book series starts with fundamental knowledge in finance, technology, quantitative methods, and financial innovation to lay the foundation for the fundamentals of fintech, and understanding the trending issues related to fintech such as regulation, applications, and global trends. It is a good starting point to the fintech literature and is especially useful for people who aspire to become fintech professionals.

Published:

Vol. 1   *Foundations for Fintech*
edited by David LEE Kuo Chuen, Joseph LIM, PHOON Kok Fai
and WANG Yu

Forthcoming:

Finance for Fintech Professionals
David Lee Kuo Chuen (Global Fintech Institute, Singapore & Singapore University of Social Sciences, Singapore), Joseph Lim, Phoon Kok Fai and Wang Yu (Singapore University of Social Sciences, Singapore)

Fintech for Finance Professionals
David Lee Kuo Chuen (Global Fintech Institute, Singapore & Singapore University of Social Sciences, Singapore), Joseph Lim, Phoon Kok Fai and Wang Yu (Singapore University of Social Sciences, Singapore)

Applications and Trends in Fintech I: Governance, AI, and Blockchain Design Thinking
David Lee Kuo Chuen (Global Fintech Institute, Singapore & Singapore University of Social Sciences, Singapore), Joseph Lim, Phoon Kok Fai and Wang Yu (Singapore University of Social Sciences, Singapore)

More information on this series can also be found at https://www.worldscientific.com/series/gfiwssf

(Continued at end of book)

Global Fintech Institute - World Scientific
Series on Fintech : 1

Foundations for Fintech

Editors

David LEE Kuo Chuen
Global Fintech Institute, Singapore
Singapore University of Social Sciences, Singapore

Joseph LIM
Singapore University of Social Sciences, Singapore

PHOON Kok Fai
Singapore University of Social Sciences, Singapore

WANG Yu
Singapore University of Social Sciences, Singapore

Published by

World Scientific Publishing Co. Pte. Ltd.

5 Toh Tuck Link, Singapore 596224

USA office: 27 Warren Street, Suite 401-402, Hackensack, NJ 07601

UK office: 57 Shelton Street, Covent Garden, London WC2H 9HE

and

Global Fintech Institute Ltd.
80 Robinson Road, #08-01, Singapore 068898

Library of Congress Cataloging-in-Publication Data

Names: Lee, David (David Kuo Chuen) editor. | Lim, Joseph, editor. |
    Phoon, Kok Fai, editor. | Wang, Yu, editor.
Title: Foundations for fintech / editors, David Kuo Chuen Lee, Global Fintech Institute, Singapore
    Singapore University of Social Sciences, Singapore, Joseph Lim, Singapore University of
    Social Sciences, Singapore, Kok Fai Phoon, Singapore University of Social Sciences, Singapore,
    Yu Wang, Singapore University of Social Sciences, Singapore.
Description: Hackensack, NJ : World Scientific, [2022] | Series: Global fintech institute -
    world scientific series on fintech; vol 1 | Includes bibliographical references.
Identifiers: LCCN 2021038693 | ISBN 9789811238802 (hardcover) |
    ISBN 9789811239267 (paperback) | ISBN 9789811238819 (ebook) |
    ISBN 9789811238826 (ebook other)
Subjects: LCSH: Financial services industry--Information technology. | Banks and banking--
    Technological innovations. | Corporate governance--Moral and ethical aspects. |
    Artificial intelligence--Economic aspects. | Risk management.
Classification: LCC HG173 .F678 2022 | DDC 332.10285--dc23
LC record available at https://lccn.loc.gov/2021038693

British Library Cataloguing-in-Publication Data
A catalogue record for this book is available from the British Library.

For any available supplementary material, please visit
https://www.worldscientific.com/worldscibooks/10.1142/12330#t=suppl

Desk Editors: Balasubramanian Shanmugam/Yulin Jiang

Typeset by Stallion Press
Email: enquiries@stallionpress.com

Printed in Singapore

Preface

The business and economic environment has been and is expected to undergo significant changes. Finance, which is both the head and heart of the flows supporting commerce, is at the forefront of such changes. And technology has enabled sea changes in how transactions are facilitated and expedited, and how financial intermediation is disrupted. Technology has further made funds available to the underserved and allocated those funds better as well as enhanced liquidity in many cases.

This interface between finance and technology requires a group of professionals who understand how finance and key technologies can be applied to facilitate financial services, and in some cases, even to disrupt traditional ways in which such services are provided. Fintech is not merely a combination of finance and technology. While digitalization is about creating new business models with digitization, fintech goes further to serve new customers with new products. To facilitate their learning journey, such professionals who straddle the finance and technology domains and occupy the intersection of both need a framework of important concepts and knowledge of both domains to understand how concepts can be put into practice. The GFI-World Scientific Series in Fintech aims to help the reader develop an understanding of fintech and its applications in an orderly fashion.

The Foundation Module

Every journey begins with the first step. We have identified key foundation knowledge underlying the science and art of fintech. In the foundation module we start in Part I with a discussion of ethics and governance where compliance with ethical principles and regulation is a necessary requirement for one to be a competent professional. Part I comprises two chapters.

Part II of the foundation module provides an understanding of the key concepts of statistical analysis and discuss them over four chapters: statistical distribution, sampling and estimation, hypothesis testing, and regression analysis.

Part III of the foundation module discusses rudiments of quantitative methods that facilitate understanding of computing and other data analysis applications. This is done over five chapters: Boolean algebra and logic gates, number system, modular arithmetic, matrix operation, and cluster analysis.

In the last two parts, key fundamentals of fintech are discussed at a very rudimentary level, where key definitions are provided, key concepts are explained, and key applications are discussed. This part introduces the basic concepts in the fundamental building blocks of fintech. Some very specific concepts in fintech that are not commonly found in finance are covered. These include the basic building blocks and ideas of decentralization, data privacy protection, peer-to-peer activities, network valuation methods, scalability, interoperability, and technical risk management. In particular, these subjects build the foundation for Level 1 subjects and to apply the concepts to actual cases in Level 2.

Part IV introduces a few important aspects of financial innovations, starting from the Fourth Industrial Revolution, followed by discussion on fintech, financial inclusion, and emerging technologies such as AI, blockchain and cryptocurrency, cloud computing, big data, and Internet of Things.

Lastly, in Part V, blockchain technology, cryptography, consensus, cryptocurrency and token economy, as well as investment related to the cryptocurrency market are further discussed.

About the Editors

Professor David LEE Kuo Chuen is a professor at the Singapore University of Social Sciences and Adjunct Professor at the National University of Singapore. He is also the founder of BlockAsset Ventures, the Chairman of Global Fintech Institute, Vice President of the Economic Society of Singapore, Co-founder of Blockchain Association of Singapore, and Council Member of British Blockchain Association. He has 20 years of experience as a CEO and an independent director of listed and tech companies and is a consultant and advisor to international organizations on food supply chain, blockchain, fintech, and digital currency.

Associate Professor Joseph LIM is with the Singapore University of Social Sciences where he teaches finance. He has also taught at the National University of Singapore and the Singapore Management University. Joseph obtained his MBA from Columbia University and PhD from New York University. In between his stints in academia, he worked in various advisory positions in the areas of private equity and valuation. Joseph, who is a CFA charter holder, has served in various committees at the CFA Institute and as President of CFA Singapore. In addition, he was on the board and committees of various investment industry associations. In the non-profit sector, he was on the board of a pension fund and several endowment funds. He is a coauthor of several popular college finance textbooks.

Associate Professor Kok Fai PHOON teaches finance at the School of Business, Singapore University of Social Sciences (SUSS). He received his PhD in finance from Northwestern University, MSc in Industrial Engineering from the National University of Singapore, and BASc in Mechanical Engineering (Honours) from the University of British Columbia. His research interests focus on the use of technology in portfolio management, wealth management, and risk and complexity of financial products. In addition to his current position at SUSS, Kok Fai has taught at other universities in Singapore and at Monash University in Australia. He has worked at Yamaichi Merchant Bank, at GIC Pte Ltd, and as Executive Director at Ferrell Asset Management, a Singapore hedge fund. He has published in academic journals including the *Review of Quantitative Finance and Accounting* and the *Pacific Basin Finance Journal* as well as practice journals like the *Financial Analyst Journal*, the *Journal of Wealth Management*, and the *Journal of Alternative Investments*.

Ms Yu WANG (Cheryl) is a fintech research fellow at the Singapore University of Social Sciences FinTech and Blockchain Group. Her main research interests are fintech, machine learning, and asset pricing. Prior to SUSS, she worked at the National University of Singapore, Business School as a research associate on corporate governance and sustainability. She graduated with an MSc in Applied Economies from Nanyang Technological University and BSc in Financial Engineering from Huazhong University of Sciences of Technologies. She has multiple journal papers, including an empirical study on sustainability reporting and firm value published on an SSCI journal, *Sustainability*, and one investigating cryptocurrency as a new alternative investment published on *Journal of Alternative Investments* that has been cited for over 200 times and recommended by *CFA Institute Journal Review.* She also serves as referee for various journals such as *Singapore Economic Review, Quarterly Review of Economics and Finance,* and *Journal of Alternative Investments.*

Contents

PART I
Ethics and Governance

Chapter 1

Ethics and Governance

Ethics and governance is an area which few of you know much. It is not taught in school and much of it is learned on the job. However, this area is one which you should take seriously. Compliance with the ethical codes and governance requirements is necessary for one to be a competent professional. Violation of the codes and governance requirements has grave consequences for the professional's reputation and ability to practice. Even if a violation does not result in some penalty, trust in the professional may be eroded and her professional standing diminished.

The scope of this module is very wide. To keep the material manageable we have focused on the fundamentals. Candidates interested in particular areas or topics can use the references to expand their knowledge and understanding.

1.1 Introduction

Ethics and governance are crucially important in corporate affairs. Any violations of proper ethical conduct or governance procedures can have adverse implications for the firm as such breaches can be easily revealed through social media. Yet, this important area is seldom taught in schools. A professional, such as the Chartered Fintech Professional (CFtP) charter holder, cannot plead ignorance of ethical and governance concerns. For this reason, the CFtP strongly emphasizes a thorough knowledge of the issues as well as a good understanding to apply the ethical and governance principles appropriately.

In the context of the CFtP curriculum, ethics and governance is about behavior, as well as incentives, whether at the personal or corporate realms. We can observe the behavior but may not know the motivation behind the action. Ethics is about ensuring that the actions undertaken by the professional have the correct motivation.

What is governance? It is a framework of policy and procedures that a company establishes to govern the conduct of corporate affairs so that actions were taken by the company and its employees are aligned to the goals of the company, comply with the laws and regulations of the state, and do not subject the company to unnecessary risk. Failure to observe corporate governance requirements can potentially have adverse outcomes leading to legal issues, financial harm, and loss of reputation and goodwill.

While poor governance could result in negative consequences, good governance can provide benefits through inspiring stakeholders' trust and confidence in the firm's policies and decisions, and yielding dividends through lower financing and transaction costs. The firm's reputation for good governance can also give it a competitive edge in securing business.

1.2　Ethical Framework

1.2.1　*Introduction*

What are the hallmarks of a true Fintech professional? (Note: For the rest of this module, the terms professional and Fintech professional are used interchangeably.) Ethical behavior and competence.

Ethical behavior inspires trust and confidence in the public that the professional will render her service with the utmost integrity. Competence gives assurance that the work or service is rendered to the highest standards of practice in the profession.

We lay the foundation for a discussion on ethics by providing a framework to think about the issues and challenges confronting ethical behavior. Then we discuss various ethical principles which can help guide us in making ethical decisions. Behaving ethically comes from the belief of the professional in doing the "right thing", which benefits her clients, her coworkers, and society at large. Integrity is infused in all aspects of the professional's work.

Beyond behaving ethically, the fintech professional has also to live up to the expectations of the public regarding professional conduct. This pertains to the quality of work and services she renders. How a fintech professional conducts herself not only projects competence but also promotes trust from the public. It is always good for the Fintech professional to benchmark her work against best practices as she is often the best critic of her professional performance.

What is ethical behavior depends on the ethical standards or principles through which we judge behavior. For our purposes, the behavior is confined to the professional realm. How a person behaves outside the work environment can have implications on professional reputation. However, issues of behavior in a non-professional setting go beyond the scope of how we expect professionals to behave, as they may entail moral and other value judgments.

Ethical judgments can be difficult and complex. Knowledge about ethics is not enough. What we need is a framework for decision-making that presents systematically a process for incorporating ethical issues in decision-making.

1.2.2 Learning Objectives

- Understand the ethical framework.

1.2.3 Main Takeaways

1.2.3.1 Main Points

- Recognizing that there are ethical issues in a situation requires familiarity with ethical principles and their application to real-world situations.
- Getting the relevant facts is crucial to making an informed decision.
- One should not be wedded to a particular course of action but should be open to alternative courses.
- Test the decision but consider how fellow professionals would view it.
- Implementing the decision must be done thoughtfully.

1.2.4 *Framework for Ethical Decision-Making*[1]

There are five steps in this framework:

1. Recognize an ethical issue
2. Get the facts
3. Evaluate alternative actions
4. Make a decision and test it
5. Act and reflect on the outcome

1.2.4.1 *Recognize an Ethical Issue*

Recognizing that there could be an ethical issue in a situation requires that the professional be familiar with ethical principles as well as how they are applied to various situations. While one may learn the principles in courses or programs like the CFtP, the facility in applying the principles can only come through practice. Learning vicariously through case studies allows the professional to encounter more situations and improve the ability to identify a potential ethical issue early. As in any skill, lack of usage will dull the skill. Therefore, ethics will be a regular feature of CFtP continuing education.

The main difficulty for this first step in the framework is that ethical issues go beyond legal and regulatory stipulations and prohibitions. Laws and regulations make clear what cannot be done or what must be done. But the demarcation between what is ethical and what is not can sometimes not be clear-cut. One way to resolve this is to consider the following:

- Could the decision or situation be disadvantageous to another party?
- What are the possible choices? Is it between a good and a bad choice, or between two "bad" or two "good" choices?
- Would the situation or issue go beyond the legal requirements?
- Is it a question of efficiency?

---

[1]This section is adapted from "A Framework for Ethical Decision Making" developed by the Markkula Center for Applied Ethics at Santa Clara University.

1.2.4.2 *Get the Facts*

Relevant facts are important inputs for analyzing a situation or issue. Decisions based on facts tend to have greater objectivity, credibility, and acceptance by the parties to the decision. Some of the questions that need to be asked are:

- Are the facts sufficient to make a decision? What more needs to be known?
- What are the interests of the parties to the situation or issue? Are certain interests more important?
- What are the options for acting? Have the issues been discussed with all the relevant parties?

1.2.4.3 *Evaluate Alternative Actions*

After marshaling the facts, the professional has to come up with various proposals to resolve the issue or situation. It is important for the professional not to be wedded to a particular option, but be open to alternatives. In proposing the alternatives the following bear consideration:

- Which option is fair to the parties?
- What is the trade-off, if any, between doing good and harm?
- Does the option chosen consider other stakeholders beyond the parties directly involved in the situation or issue?

1.2.4.4 *Make a Decision and Test It*

Selecting the best course of action is not easy as it is often not obvious which option is best. Whichever the option that best resolves the situation or issue at hand, the chosen option should not conflict with ethical principles. A good way to test the soundness of the decision is to consider how your fellow professionals would view the option you have chosen.

1.2.4.5 *Act and Reflect on the Outcome*

Making a good decision but implementing it poorly may not lead to the desired effect. The concerns of the stakeholders must be taken into account as ultimately it is their perception of fairness that determines, to a large extent, whether the implementation is satisfactory.

It is always good for the professional to reflect on the decision taken and the outcome of the decision so that lessons can be drawn from the experience. Ideally, the professional does not keep the lessons learned to herself but endeavors to share them with the wider body of professionals. This would help the profession progress to higher standards of ethical practice and justify the public's trust in them.

1.3   Ethical Principles: CFtP Code of Ethics

1.3.1   *Introduction*

The CFtP Code of Ethics is a set of ethical principles that all CFtP charter holders must uphold. These principles inform the charter holders' decisions when they have to make difficult decisions when conflicting interests are involved. Making the right decision calls for a solid knowledge of the principles as well as the skill to apply them in complex situations.

1.3.2   *Learning Objectives*

• Discuss the four ethical principles and their implications.

1.3.3   *Main Takeaways*

1.3.3.1   *Main Points*

• Integrity is the foundation and basis for trust in the work of a professional.
• Fairness requires that the actions and decisions of the professional are inclusive and not restricted to a favored few.
• In performing their work, professionals should exercise the duty of care to avoid adverse outcomes due to oversight.
• Professionals who respect the rights of others do not abuse their advantages due to their position or access to privileged information.

1.3.4 *Integrity*

1.3.4.1 *Honest and Trustworthy*

A professional has integrity when she acts honestly, is trustworthy, and is guided by strong morals of right and wrong.[2] Her behavior is consistent and she is not easily swayed by appeals, arguments, or threats antithetical to her ethical values causing her to be viewed as a trustworthy person.

A professional possesses knowledge and skills that often put her in an advantageous position over her clients. To ensure that she works and renders service with integrity, the professional should "place the integrity of the profession and the interests of the clients above their interests".[3] In this regard, the professional should disclose any conflicts of interest "that could reasonably be expected to impair their independence and objectivity or interfere with respective duties to their clients, prospective clients, and employer".[4]

1.3.5 *Fairness*

1.3.5.1 *No Discrimination*

The principle of fairness incorporates "values of equality, tolerance, respect for others and justice".[5] This principle requires that the professional embraces inclusiveness and does not discriminate, putting aside her personal preferences and predispositions. Concerning clients, the professional must "deal fairly and objectively ... in professional activities".[6]

Fintech professionals should be mindful of the impact of information and technology either creating or perpetuating inequities.

---

[2] ACM Code of Ethics and Professional Conduct section on General Ethical Principles, Clause 1.3.

[3] CFA Institute Code of Ethics and Professional Conduct: The Code of Ethics.

[4] CFA Institute Standards of Professional Conduct: Conflicts of Interest, A. Disclosure of Conflicts.

[5] ACM Code of Ethics and Professional Conduct section on General Ethical Principles, Clause 1.4.

[6] CFA Institute Standards of Professional Conduct: Duties to Clients, Clause B. Fair Dealing.

The design of systems should factor in inclusiveness and not allow them to be used for purposes of discrimination.

1.3.6 Duty of Care

1.3.6.1 Avoiding Harm

1.3.6.1.1 Negative Consequences

In the performance of their work, the professional should exercise the duty of care. It is expected that a professional is competent and the work rendered is of a high standard. By not exercising care in the course of fulfilling the work, the professional may end up causing harm.

1.3.6.1.2 Indirect or Unintentional Consequences

The failure to exercise care can have indirect and unintentional consequences. Negligence on the part of the professional can have a direct cost, for example, the loss of a client. The indirect cost is the tarnishing of the firm's reputation.

1.3.6.2 Duty to Report Signs of Failure or Risks that May Cause Harm

Finally, the duty of care imposes on the professional to report signs of failure or risks that may cause harm.[7] A professional is well-trained and has the knowledge and skills to spot potential risks and failures. Her expertise would serve the firm well as the typical firm usually has a small number of risk specialists.

1.3.7 Respect for the Rights of Others

1.3.7.1 Respect Privacy

An important part of respecting the rights of others is when you respect their privacy. This is highly relevant today when "technology

---

[7]ACM Code of Ethics and Professional Conduct section on General Ethical Principles, Clause 1.2.

enables the collection, monitoring, and exchange of personal information . . . often without the knowledge of people affected".[8]

1.3.7.2 *Respect Intellectual Property and the Contributions of Others*

Going beyond the personal realm is the respect for the intellectual property and contributions of others. This means that professionals should "credit the creators of ideas, inventions, work, and artifacts, and respect copyrights, patents, trade secrets, license agreements, and other methods of protecting authors' works".[9]

1.3.7.3 *Safeguard Confidentiality*

Having access to private information and put in a position of trust, the profession is required to safeguard the confidentiality of the information she is privy to. An exception to maintaining confidentiality is any requirement to disclose to the appropriate authorities.

1.3.7.4 *Respecting the Concerns of the Society*

Beyond the professional's immediate sphere of influence, there is a need to be aware of the society at large. The concept of sustainability has taken root in the corporate agenda and respecting the rights of others extends to other stakeholders, whether direct or indirect. For example, the Singapore Exchange introduced sustainability reporting on a "comply or explain" basis in 2016 for companies listed on their exchange.

1.4 Professional Practice: CFtP Standards of Practice

1.4.1 *Introduction*

Professional practice is concerned about how the professional should go about doing her work. A foundation for good professional practice is competence. Without competence, the standard of work may not be at a level commensurate with the high standards expected of a

---

[8] *Ibid*, Clause 1.6.
[9] *Ibid*, Clause 1.5.

professional. As the status of a profession is often constituted by the authorities, it is incumbent upon the professional be conversant with the laws and regulations governing her area of work. Further, professionals often deal with sensitive information, some of which are personal. Protecting the information and maintaining confidentiality is another aspect of professional practice. Finally, an area that is not often brought up in discussing professional practice is leadership. As professionals are held in high regard, they are often called upon to exercise leadership at work and in the society at large, and hence the responsibility that leadership entails has to be highlighted.

1.4.2   *Learning Objectives*

- Discuss the CFtP Standards of Practice and their implications.

1.4.3   *Main Takeaways*

1.4.3.1   *Main Points*

- Competence requires knowledge and skills guided by ethical values to achieve high-quality work.
- The professional should be mindful of not doing work outside their realm of expertise.
- The professional should ensure that those performing the work have the competence to achieve a high standard of work.
- Knowledge of applicable laws and regulations is essential to ensure her compliance.
- Through their work, professionals may have sensitive information which they are bound to keep confidential as well as ensuring the information is kept secure.
- As professionals are responsible for those under their charge, they are to exercise leadership in three areas: the work itself, the welfare and dignity of subordinates, and the implication for the social good of the professional work.
- Upholding the reputation of the CFtP designation is the responsibility of all charter holders.

1.4.4   *Competence*

Competent professionals are those who possess the knowledge, skills, and ethical values to "achieve high-quality in both the processes and the products of professional work".[10] They "[m]aintain and improve their professional competence and strive to maintain and improve the competence of their [fellow] professionals".[11]

1.4.4.1 *Knowledge*

Knowledge is obtained through a course of study in school or through a certification program like the CFtP. Also, the professional needs to update her knowledge through continuing education given the quick pace of change taking place in the world as a result of new technologies and innovations. The CFtP requires that charter holders continually upgrade their knowledge through its continuing education programs.

1.4.4.2 *Skills*

The skills that the professional possesses come from practice in applying the knowledge acquired to solve problems in the professional's sphere of expertise. Skills can be enhanced through interaction with fellow professionals at conferences and meetings as well as through case studies.

1.4.4.3 *Ethical Values*

Ethical behavior is discussed extensively in the ethics section of this module. Ethical behavior cannot be overemphasized. Unless the professional is guided by ethical values in using her knowledge and skills in her practice, she cannot be considered a true professional.

1.4.4.4 *Area of Work*

CFtP members are also admonished to perform work only in their areas of competence. They are to "[evaluate] the work's feasibility and advisability, and [make] a judgment about whether the work assignment is within the professional's areas of competence. If at any time before or during the work assignment the professional identifies

---

[10] ACM Code of Ethics and Professional Conduct section on Professional Responsibilities, Clause 2.1.
[11] CFA Institute Code of Ethics and Standards of Professional Conduct: The Code of Ethics.

a lack of necessary expertise, they must disclose this to the employer or client".[12]

1.4.5 Standard of Work

High standards of work can be achieved through individuals and their teams taking responsibility to ensure that they are competent.[13] However, competence is just a prerequisite of high-quality work. The professional must be guided by the ethical principles of integrity, fairness, and duty of care toward her clients and employer. It is all too easy to produce a quality of work that is satisfactory given the constraints of time and considerations of profitability. Giving your best comes from not just a duty of responsibility but a sense of professional pride.

1.4.6 Knowledge of Law and Regulations

In addition to domain knowledge, professionalism requires a knowledge of the law and regulations that govern the sphere of the professional's work. The professional "must understand and comply with all applicable laws, rules, and regulations ... [including the CFtP Code of Ethics and Professional Practice] of any government, regulatory organization, licensing agency, or professional association governing their professional activities. In the event of a conflict, [members]...must comply with the more strict law, rule or regulation".[14]

1.4.7 Confidentiality and Security

The professional should be aware of the importance of ensuring proper security over the intellectual property and confidentiality of

---

[12] ACM Code of Ethics and Professional Conduct section on Professional Responsibilities, Clause 2.6.

[13] *Ibid*, Clause 2.2.

[14] CFA Institute Code of Ethics and Standards of Professional Conduct: Standards of Professional Conduct, Section I Professionalism, A. Knowledge of the Law, as well as ACM Code of Ethics and Professional Conduct section on Professional Responsibilities, Clause 2.3.

her work.[15] Any breach can be costly as the professional and her employer may be sanctioned by the authorities or face lawsuits from affected parties. Further, the damage from such a breach could have ramifications on the business of the professional. The affected party's competitors may gain an edge from the confidential information released and the firm's reputation would be tarnished, leading to some existing clients leaving due to a lack of confidence and trust in the company. Such adverse publicity could also make it difficult to acquire new clients.

Data breaches could violate data protection laws. In the age of the digital and information economy, data is an extremely important resource and should be well protected. This means that "[r]obust security should be a primary consideration when designing and implementing systems ... [and] parties affected by data breaches are notified in a timely and clear manner providing appropriate guidance and remediations".[16] In dealing with clients there is a need to "keep information about current, former and prospective clients confidential unless the information concerns illegal activities, [or] disclosure is required by law, or the client, or prospective client permits".[17]

1.4.8 *Leadership*

Beyond the technical realm of work, the professional is also called upon to provide leadership in her organization and the community-at-large. The areas and circumstances where leadership should be exercised are articulated in some detail in the section on professional leadership principles in the ACM Code of Ethics and Professional Conduct. We highlight a few principles here.

---

[15] ACM Code of Ethics and Professional Conduct section on General Ethical Principles, Clause 1.7.

[16] ACM Code of Ethics and Professional Conduct section on Professional Responsibilities, Clause 2.9.

[17] CFA Institute Standards of Professional Conduct: Duties to Clients, Clause E. Preservation of Confidentiality.

1.4.8.1   *Management of Personnel*

1.4.8.1.1   Quality of Working Life and Promotion of Professional Growth

Work–life balance and professional development of the employee are increasing concerns which employees expect their employers to provide. "Leaders should enhance, not degrade, the quality of working life ... [and] consider the personal and professional development ... and dignity of all workers".[18]

1.4.8.1.2   Social Responsibilities

As a leader, the professional has to ensure that "the public good is the central concern during all professional work"[19] as well as, for the IT professional, "recognize and take special care of systems that become integrated into the infrastructure of society".[20]

1.4.9   *Upholding the Reputation of the CFtP Designation and the Profession*

The CFtP is a designation that members expend much effort to secure. Any untoward action of a member has the potential to tarnish the image and reputation of the CFtP as well as the reputation of the profession in general. It is therefore incumbent on every member to safeguard this reputation as well as report any member whom they have reason to believe is acting in a manner contrary to the Code of Ethics and Professional Practice which could harm the reputation of the CFtP designation as well as the Global Fintech Institute, the awarding body.

1.5   Summary

1.5.1   *Fintech Ethical Issues*

The requirement for the fintech professional to behave ethically may not appear too different from the requirements of other professions.

---

[18] *Ibid*, Clauses 3.3 and 3.5.
[19] *Ibid*, Clause 3.1.
[20] *Ibid*, Clause 3.7.

However, there is one important distinction. In fintech, the technology involved can have far-reaching ramifications. The issues with digital technology are well known. The Level 2 CFtP curriculum discusses some of the issues extensively. In this summary, we highlight some of them.

In the area of artificial intelligence (AI), the issues go beyond machines replacing humans in many job functions. Of particular concern is the delegation of decision-making to AI. How far should it go? Who would be held responsible for bad decisions? Is it the software developer? Is it the consultant who helps implement the software? Is it the user?

Given the anonymous nature of Blockchain transactions, and the lack of a central coordinating entity, who is responsible for abuses of this feature of Blockchain? Should the use of the Blockchain to launder money or to engage in illegal activities curtail its use? Governments in their attempts to control Blockchain activities have occasionally used blunt regulatory instruments such as an outright ban. How should the fintech professional finesse the issue of good vs harm?

Cybersecurity is a very serious problem for security experts. Unlike physical installations where access cannot be done remotely, the very nature of the digital world allows intrusions from anywhere in the world. Much loss and harm can occur when a breach of cybersecurity occurs. The Fintech professional has to be well aware of security concerns.

Finally, personal data protection has become another prominent concern with issues like identity theft, misuse of personal data, and exploitation of vulnerable populations. The Fintech professional often has to grapple with the trade-off between security and efficiency.

How is the fintech professional going to navigate the complex cyber world that is ever-changing and make good ethical decisions? We reiterate some of the ethical principles and provide some guidelines.

1.5.2 *Be Grounded in Ethical Principles*

1. Always act with integrity

   - Do not put self-interest first

2. Be fair in your actions

   • Do not discriminate, practice inclusiveness

3. Exercise duty of care

   • Avoid doing harm
   • Be aware of negative or unintended consequences

4. Respect the rights of others

   • Respect privacy and safeguard the confidentiality

1.5.3   *Resolving Fintech Ethical Issues*

Fintech is responsible for many disruptions in the business and financial world as well as in ordinary life. How do we view nascent technology that has the potential to disrupt the status quo?

To make an informed decision there needs to be sufficient knowledge grounded in evidence and accepted precedents. Advisers and experts in the field can provide perspective. Another source is crowdsourcing. It is good to seek a wide variety of views as one is exercising the duty of care. The fintech professional should not adopt the attitude of being agnostic regarding technology.

While technology and innovation hold the promise of improvements, the fintech professional should be aware of negative or unintended consequences. If the introduction of technology does more harm than good, then it should not be introduced at all. However, what do we mean by good or harm?

Technology is viewed positively if it results in the betterment of society. For example, if the technology results in greater efficiency, then it is good. However, if the technology disadvantages certain groups of people do we consider it harmful? If one group is the incumbents and the new technology results in this group having a competitive disadvantage, this cannot be considered a harm. With innovation, there will always be winners and losers. Examples are the demise of the horse and buggy industry upon the arrival of the automobile, or steam engine engineers whose skills became somewhat obsolete with the advent of the internal combustion engine.

What about harm? The introduction of nuclear reactors cut down the emissions produced by power stations burning fossil fuels. This "good" however came with an unintended "harm". The radiation leak from the Fukushima nuclear reactor in Japan in 2011 caused a large swathe of land around it to be uninhabitable for a considerable length of time. The discharge of radioactive cooling water into the sea resulted in contaminated fish. How do we resolve this difficult problem?

The fintech professional does not need to grapple with the issues alone. Governing authorities and the regulators confront competing interests from their stakeholders. They would be in the best position to make a judgment call as politics is about resolving issues affecting society. After all elected officials represent the aspirations and wishes of the majority of society.

Bibliography

As there are many topics in this module and each topic, in itself, has a wide scope, we are only able to discuss the fundamentals. The references here provide the candidates with material that are relevant to their areas of work and go into greater depth as well as cover the applications, laws, and regulations in various jurisdictions.

Ethical Framework and Principles

Association for Computing Machinery (2018). ACM code of ethics and professional conduct. Retrieved from https://www.acm.org/code-of-et hics.

CFA Institute (2014). Code of ethics and standards of professional conduct. Retrieved from https://www.cfainstitute.org/-/media/documents/cod e/code-ethics-standards/code-of-ethics-standards-professional-conduc t.ashx.

IESBA (2019). International code of ethics for professional accountants. Retrieved from https://www.ethicsboard.org/international-code-ethic s-professional-accountants.

Indeed (2021). 10 characteristics of professionalism. Retrieved from https://www.indeed.com/career-advice/career-development/the-ulti mate-guide-to-professionalism.

Markkula Center for Applied Ethics at Santa Clara University (2015). A framework for ethical decision-making. Retrieved from https://www. scu.edu/ethics/ethics-resources/ethical-decision-making/a-framework -for-ethical-decision-making/.

National Society of Professional Engineers (NSPE) (2019). NSPE code of ethics for engineers. Retrieved from https://www.nspe.org/resources/ ethics/code-ethics.

Vivian, W. (2008). Illinois institute of technology center for the study of ethics in the professions. Retrieved from http://ethics.iit.edu/teaching /professional-ethics.

1.6   Sample Questions

Question 1

Who is least likely to be required to observe the CFtP Code of Ethics and Standards of Practice?

(a) CFtP exam candidate.
(b) Associate member of GFI
(c) Subordinate of CFtP charter holder

Question 2

Which is the least likely desirable conduct for CFtP exam candidates?

(a) Acting with integrity when dealing with competitors
(b) Being transparent with clients
(c) Balancing the interests of the employer with the need in the profession to act with integrity

Question 3

Tom, a CFtP charter holder, was in charge of multiple projects. For one of the projects, he needed to make a filing with the authorities to satisfy a regulatory requirement. There was no time to consult the expert in this area, and Tom who has worked in this area before decided to do the filing himself. Unfortunately, Tom did not know that the regulation had been changed and the firm had to pay a penalty for making the wrong filing. Tom is guilty of violating which CFtP Standard of Practice

(a) Competence and leadership
(b) Competence and knowledge of laws and regulations
(c) Leadership and knowledge of laws and regulations

Question 4

Many of the staff working under John, a CFtP charter holder, complained about having to work till late evening every day as well as having to work every weekend. Work–life balance is:

(a) the firm's responsibility
(b) the HR department's responsibility
(c) John's responsibility in his exercise of leadership

Solutions

Question 1

Solution: Option **c** is correct.

The CFtP charter holder has to exercise leadership in the organization beyond her professional or job scope to foster ethical behavior. However, the subordinate, unlike the other two persons, has no direct obligation to observe the CFtP Code of Ethics and Standards of Practice.

Question 2

Solution: Option **c** is correct.

While the CFtP exam candidate should work for the best interests of the employer, this fact should not compromise the need to act with integrity.

Question 3

Solution: Option **b** is correct.

Tom is acting beyond his level of competence as he is not an expert in the area. He is also guilty of not keeping up with the laws and regulations.

Question 4

Solution: Option **c** is correct.

Under the CFtP Standard of Practice, it is stipulated under the standard on leadership that the CFtP charter holder should consider the work-life balance of his subordinates in his exercise of leadership.

Chapter 2

Governance

2.1 Introduction

Regulation is an intervention by the state in a society to require or proscribe conduct. In a society where competing interests exist, the greater good of the society may be affected by the actions of some, whether inadvertently or not. By laying down the rules governing the transactions of corporations and individuals, the economy and society can function more efficiently in catering to the needs of its members.

As the interests of the participants in an economy are diverse, it is impossible to satisfy the needs of everyone. Recognizing this fact, it is evident that regulation is a compromise that serves to ensure that the greater good of society is a key consideration.

Professionals have a role in examining regulations in place or regulations in the making to ensure that they are relevant to society and would benefit society at large. Hence, a knowledge of the aims of regulation as well as the considerations regulators has to take into account in drafting the regulations is important. It is only when professionals understand the regulatory process that they can then set about to comment on and influence policies from a perspective that draws on their specialized knowledge, skills, and ethical values.

What about governance? How different is governance from regulation? A simple answer is that governance is much broader than regulation. Given their nature, the sword of laws and regulations has to be sharper than that of governance. For laws and regulations, it is a matter of whether they have been violated or not. On the other hand, governance is sometimes a matter of degree. We talk of good vs poor governance.

2.2 Governance and Regulatory Framework

2.2.1 *Introduction*

The regulation of fintech consists of two strands: regulation of financial activities and regulation of IT activities. To some extent, the two strands operate separately and the legacy of existing regulations results in each strand seldom considering implications of or for the other strand. However, as fintech becomes more embedded in the economy, regulations may have to adopt a more holistic approach. This has been seen, for example, in cryptocurrency and payment systems.

For the CFtP program, a holistic approach to regulation will not be attempted. The reason is that many fintech activities are recent developments and few regulatory and legal issues have been tested in the courts or a strong consensus among stakeholders of how some of those issues ought to be settled has not evolved. Hence, no attempt will be made to unify both strands of regulation. Each would be discussed separately. Further, as local regulations take into account each jurisdiction's unique circumstances and approach, the examples that are cited are meant for illustrative purposes, showing how various jurisdictions have chosen to address some of the issues on that regulation.

2.2.2 *Learning Objectives*

- Understand the key elements of the framework for regulation and governance.

2.2.3 *Main Takeaways*

2.2.3.1 *Main Points*

- The three components of regulation are incentive, deterrence, and enforcement.

- The principle of incentive is to promote desired behavior.
- Enforcement of laws and regulations are less effective or not effective at all outside a country's jurisdiction. The internet is a good example.
- Many cross-border transactions are governed by uniform codes or international conventions.
- Where regulatory regimes have little effect because cross-border transactions are digital, and enforcement is difficult, self-regulation is the only alternative.

2.2.4 General Framework

Regulation relies on the three components of incentive, deterrence, and enforcement.

The principle of incentive aims to promote desired behavior while the principle of deterrence aims to deter undesired behavior. However, regulation is ineffective if it is not backed by enforcement through sanctions or penalties against undesired behavior.

Enforcement is effective only within a particular jurisdiction where the authorities have legal sway over people and entities domiciled in that jurisdiction. In a globalized world where cross-border transactions are common, the issue of enforcement takes on another dimension as the arm of the law is not long enough to extend beyond the national borders. Introduce the world of the internet and enforcement becomes even more complicated.

Many cross-border transactions in assets, goods, or services may be governed by some international convention that counts most countries in the world as signatories. An example is the United Nations Convention on Contracts for the International Sale of Goods, which provides a framework for governing international commerce. Where the cross-border transaction is not governed by some uniform code or convention, the parties involved need to decide on the issue of the applicable governing law and dispute settlement mechanisms.

Where cross-border transactions are digital, regulation is challenging because of the difficulty of enforcement. When regulatory regimes break down, the only effective regulation is self-regulation. This requires a focus on the incentive. Some of the developments in the area of digital regulation embed incentive mechanisms in their design.

In the earlier sections on the CFtP Code of Ethics and Standards of Practice, we note that violation of the code and standards may bring forth sanctions from the professional body. The power of the professional body to sanction a member is quite limited. The most severe sanction is expulsion from the professional body. Unless the professional's ability to practice her specialty is contingent on the person being a member of a professional body, expulsion may not have much force to ensure compliance.

Regulations and laws allow the enforcing authority to impose penalties much more severe than those meted out by a professional body. That said, it does not mean the professional should view the observance of the code of ethics and standards of conduct lightly. Rather, if the professional embraces both the code and the standards and observes them strictly, it is highly unlikely such a professional would run afoul of the law.

2.2.4.1 *What is Being Governed?*

Unlike conventional governance where humans verify compliance, the new world of fintech introduces a new mode of verification. Technology, especially the blockchain, can be used in the verification process. Not requiring the active hand of human verification, technology is able to streamline the verification process and make it more efficient. An added benefit of monitoring compliance through technology is that verification can be done on a real-time basis. This has tremendous benefits as malfeasance can be spotted early and corrective action taken before much damage can be done.

An illustration of how technology adds a new dimension to verification of compliance is shown in Figure 2.1. A more detailed discussion of this aspect will be presented in the CFtP Level 2 curriculum.

2.3 Governance and Regulatory Principles

Some of the governance and regulatory principles may overlap with the ethical principles on which we base the Code of Ethics and Standards of Practice. Examples of the overlap would be the principle of fairness and the principle of respect for the rights of others. Where the principles have been discussed in the Ethics section, we will only discuss their application from the regulatory viewpoint.

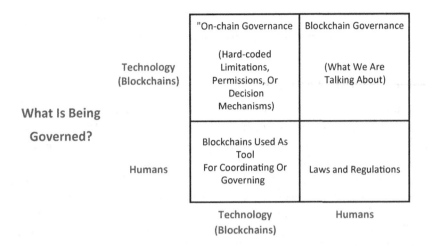

Figure 2.1: Technology and compliance.

2.3.1 *Learning Objectives*

• Discuss the general governance and regulatory principles.

2.3.2 *Main Takeaways*

2.3.2.1 *Main Points*

• Regulation and governance require that professionals are accountable for performing actions required by regulation as well as refraining from actions prohibited by regulation.
• In the context of IT, equity or fairness requires that the professional consider the potential impact of her work on the wider society, that is, to think beyond the confines of their organization or even the country she works in.
• Showing transparency in the software gains the trust of users.

2.3.3 *Accountability*

The principle of accountability rests on the notion that an individual is responsible for the performance of certain actions as prescribed by the regulation as well as refraining from performing actions prohibited by the regulation. However, accountability goes beyond the individual's action. Where the professional has people under her charge, she is also accountable for their observance of the regulation.

2.3.4 Equity/Fairness

Fairness was discussed as one of the ethical principles those holding the CFtP charter need to uphold. In the context of IT, systems and software like AI have "pervasive, far-reaching and global implications that are transforming societies, economic sectors and the world of work". Knowing the potential impact on the wider society, fairness implies that the professional should think beyond the confines of her organization or even the country she works in.

The principle of equity means that the professional should consider the following:

- *The individual*
  Concerns of the individual encompass privacy issues as well as inclusiveness. Are the individual's rights part of the decision-making process?
  Is her privacy protected? Are individuals discriminated against because of status, disability, or income?
- *Society*
  Are there segments of the society that are left out because of their particular circumstances, for example, the less educated, the less IT-literate, or those who do not know the language in which the program is written? Do software and systems lead to the development and progress of society? Are issues of sustainability factored in?

2.3.5 Transparency

Software and systems are often perceived to be arcane black boxes. Developers hide behind a veil of secrecy to thwart the prying eyes of competitors and other parties. Consequently, users are at the mercy of bugs and hidden features.

The professional who embraces the principle of transparency will disclose as much as possible, subject to the constraint of proprietary intellectual property, to gain the trust of users. Inefficiencies due to the asymmetry of information will be greatly reduced through transparency. Open software encapsulates this principle of transparency.

2.4 Areas of Governance and Regulation

2.4.1 *Technology*

The fast pace of technological innovations presents a few problems for governance and regulation which have to play catch-up. There are several reasons for this:

a. It takes time for those in charge of governance and regulation to understand the new technology. The technical nature of some of these regulations implies a longer learning curve.
b. There may not be governance codes and regulations that can be adapted from other areas. Having to draft codes and regulation from scratch requires much discussion and consultation which may require considerable time.
c. Some implications of the new technology will only surface with more use cases.

The CFtP curriculum will cover four areas of technology, corresponding to the first four letters of the alphabet. They are AI, Blockchain, Cybersecurity, and Data Protection/Privacy. These topics will not be discussed here but in CFtP Level 2 Module 2.1 Ethics and Governance.

2.4.2 *Finance*

The finance industry is highly regulated given the potential systemic risks it poses and its importance in the national economy. Many of the fintech innovations have disrupted the finance industry whether intended or not. As fintech professionals work at the intersection of finance and technology, they need to be aware of some of the financial regulations.

Two key regulations are the Dodd–Frank Act and the MiFID II directive. The first stems from the Great Financial Crisis (GFC) of 2008. It was enacted in the United States to regulate lenders and banks to protect consumers as well as ensure the soundness of banks and financial institutions. Banks were also required to increase their regulatory capital so that the probability of bank failures is reduced. The capital that banks need to put up conform to the Basel III framework, which is risk-based where the amount of capital required depends on the quantity as well as the risk of the assets.

MiFID II, like the Dodd–Frank Act, was enacted in the aftermath of the GFC to restore confidence in the financial industry. It is a framework that covers the trading systems, the assets traded, and the profession as well. MiFID II is discussed in greater detail in Level 2 Module 2.5 on regulation.

The CFtP curriculum will focus on four aspects of governance in this section.

- Anti-Money Laundering/Combating the Financing of Terrorism (AML/CFT)
  This is highly relevant for fintech as there are many applications in payments and online investment.
- Corporate Governance
- Environmental, Social, and Corporate Governance (ESG)
  This encompasses notions of sustainability and fairness.
- Sustainability and Green Finance

These topics will not be discussed here but in CFtP Level 2 Module 2.1 Ethics and Governance.

Bibliography

As there are many topics in this module and each topic, in itself, has a wide scope, we are only able to discuss the fundamentals. The references here provide the candidates with material that are relevant to their areas of work and go into greater depth as well as cover the applications, laws, and regulations in various jurisdictions.

Technology

General

McKie, S. (2018). Blockchain communities and their emergent governance. Retrieved from https://medium.com/amentum/blockchain-communiti es-and-their-emergent-governance-cfe5627dcf52.

AI

Info-communications Media Development Authority (IMDA) and Personal Data Protection Commission (PDPC) (2020). *Model Artificial Intelligence Governance Framework.* 2nd edition. Retrieved from https://www.pdpc.gov.sg/-/media/Files/PDPC/PDF-Files/Res ource-for-Organisation/AI/SGModelAIGovFramework2.pdf.

Monetary Authority of Singapore (2018). Principles to promote fairness, ethics, accountability and transparency (FEAT) in the use of artificial intelligence and data analytics in Singapore's financial sector. Retrieved from https://www.mas.gov.sg/publications/monographs-or -information-paper/2018/FEAT.

OECD (2019). Recommendation of the council on artificial intelligence. Retrieved from https://legalinstruments.oecd.org/en/instruments/OE CD-LEGAL-0449.

World Economic Forum (WEF) (2019). AI governance: A holistic approach to implement ethics into AI. Retrieved from https://www.weforum.org/whitepapers/ai-governance-a-holistic-approach-to-implement-ethics-into-ai.

World Economic Forum (WEF) (2020). Ethics by design: An organizational approach to the responsible use of technology. Retrieved from https://www.weforum.org/whitepapers/ethics-by-design-an-org anizational-approach-to-responsible-use-of-technology.

Blockchain

Lapointe, C. and Fishbane, L. (2019). The blockchain ethical design framework. *Innov. Technol Gov Glob.*, 12(3/4), 50–71. Retrieved from https ://www.mitpressjournals.org/doi/pdf/10.1162/inov_a_00275.

Cybersecurity

International Chamber of Commerce (ICC) (2015). ICC cybersecurity guide for business. Retrieved from https://iccwbo.org/publication/icc-cyber -security-guide-for-business/.

Monetary Authority of Singapore (2021). Technology risk management guidelines. Retrieved from https://www.mas.gov.sg/-/media/MAS/R egulations-and-Financial-Stability/Regulatory-and-Supervisory-Fram ework/Risk-Management/TRM-Guidelines-18-January-2021.pdf.

National Institute of Standards and Technology (NIST) (2018). Cybersecurity framework. Retrieved from https://www.nist.gov/cyberframew ork.

Data Protection/Privacy

iSight (2021). A practical guide to data privacy laws by country. Retrieved from https://i-sight.com/resources/a-practical-guide-to-data-privacy-laws-by-country/(2021).

Personal Data Protection Commission Singapore (2012). Singapore personal data protection act. Retrieved from https://www.pdpc.gov.sg/Overview-of-PDPA/The-Legislation/Personal-Data-Protection-Act.

Finance

Anti-Money Laundering/Combating the Financing of Terrorism (AML/CFT)

AML-CFT (2017). 6 elements of an effective AML/CFT compliance programme. Retrieved from https://aml-cft.net/6-elements-effective-amlcft-compliance-programme/.

International Monetary Fund (IMF) (n.d.). Anti-money laundering/combating the financing of terrorism (AML/CFT). Retrieved from https://www.imf.org/external/np/leg/amlcft/eng/.

International Finance Corporation (IFC) (2019). Anti-money-laundering (AML) & combating the financing of terrorism (CFT) risk management in emerging market banks. Retrieved from https://www.ifc.org/wps/wcm/connect/e7e10e94-3cd8-4f4c-b6f8-1e14ea9eff80/45464_IFC_AML_Report.pdf?MOD=AJPERES&CVID=mKKNshy.

Corporate Governance

CFA Institute (2018). *The Corporate Governance of Listed Companies.* 3rd Edition, A Manual for Investors. Retrieved from https://www.cfainstitute.org/en/advocacy/policy-positions/corporate-governance-of-listed-companies-3rd-edition.

Corporate Finance Institute (CFI) (n.d.). What is corporate governance? Retrieved from https://corporatefinanceinstitute.com/resources/knowledge/other/corporate-governance/.

Harvard Business Review (2019). A guide to the big ideas and debates in corporate governance. Retrieved from https://hbr.org/2019/10/a-guide-to-the-big-ideas-and-debates-in-corporate-governance.

Environmental, Social and Governance (ESG)

CFA Institute (2015). Environmental, social and governance issues in investing – a guide for investment professionals. Retrieved from https://www.cfainstitute.org/en/advocacy/policy-positions/envi

ronmental-social-and-governance-issues-in-investing-a-guide-for-inve
stment-professionals.

McKinsey Quarterly (2019). Five ways that ESG creates value. Retrieved
from https://www.mckinsey.com/business-functions/strategy-and-cor
porate-finance/our-insights/five-ways-that-esg-creates-value.

MSCI (n.d.). ESG101: What is ESG? Retrieved from https://www.msci.c
om/what-is-esg.

Sustainability and Green Finance

Green Finance Platform (n.d.). Explore green finance. Retrieved from htt
ps://www.greenfinanceplatform.org/page/explore-green-finance.

World Economic Forum (WEF) (2020). What is green finance and why is
it important? Retrieved from https://www.weforum.org/agenda/2020
/11/what-is-green-finance/.

2.5 Sample Questions

Question 1

The foundation and basis for trust in the work environment are best
described by which of the following qualities in the professional:

(a) Compliance with laws and regulations
(b) Integrity
(c) Possession of the necessary knowledge and skills

Question 2

The professional who possesses privileged information concerning the
client should:

(a) Use it for the benefit of the firm
(b) Respect the rights of the client and not abuse the advantage
(c) Balance the interests of the employer with the interests of the
client

Question 3

Enforcement of rules and regulations:

(a) Is always possible because of the power of the governing authority
(b) Is dependent on the co-operation of the different parties
(c) Is not always possible because of cross-border issues

Question 4

Transparency in software:

(a) Is a requirement of regulations
(b) Builds trust in the user
(c) Is always there when the software does not employ a "black-box"

Solutions

Question 1

Solution: Option **b** is correct.

Compliance with laws and regulations is a requirement for all professionals. Having the knowledge and skills is a prerequisite for doing a competent job. However, trust is based on the integrity of the professional.

Question 2

Solution: Option **b** is correct.

Putting the interest of the client first is the basis of good ethical behavior. The professional has an obligation not to abuse the advantage she has from possessing privileged information about the client.

Question 3

Solution: Option **c** is correct.

Option a is not correct as governments have difficulty extending their authority beyond their borders. Option b is not correct as compliance with regulations is mandatory and not optional. Option c is correct especially when cross-border transactions are digital.

Question 4

Solution: Option **b** is correct.

Option a is not correct as regulations hardly, if ever, mandate transparency. Option c is not correct as the absence of a "black-box" does not necessarily mean the software is easily understandable. Option b is correct as users develop trust in the software as it is open.

PART II
Statistics

Chapter 3

Introduction and Probability Distribution

3.1 Introduction

The framework for statistical analysis comprises the sample space, collection of events, and a probability measure. In many applications, we are concerned about the distribution of random variables defined in the sample space. In particular, we may wish to estimate quantities such as the average value or the dispersion of the random variable about its average value. In other applications, we may wish to determine whether there is sufficient evidence to support a claim.

Estimation of quantities and testing of a hypothesis are two major endeavors of statistical analysis that are carried out using information obtained from a random sample of a population. Thus the knowledge of the sampling distributions of statistical quantities associated with a random sample is key to tackling estimation problems or hypothesis testing. An important result known as the central limit theorem states that for any distribution with finite variance, the sampling distribution of the sample mean of a random sample drawn from the distribution when properly normalized approaches the standard normal distribution.

3.2 Probability Distributions

A probability distribution is a function that describes the likelihood of obtaining the possible values that a random variable can assume.

A random variable is a real-valued function on the sample space that is of interest and it inherits a probability distribution from the probability measure. The probability distribution associated with a random variable encapsulates all relevant information about the random variable. Typically we are interested in the first and second moments, namely the mean and variance of a distribution. We review the basic properties of common distributions such as uniform distribution, binomial distribution, Poisson distribution, and normal distribution. We shall apply properties of the sampling distributions to estimate population parameters and to test a hypothesis concerning populations. Along the way, we point out the connection between confidence interval estimates of a parameter and the testing of hypotheses concerning it.

3.2.1 *Learning Objectives*

- Understand the concept of probability distribution properties and its methods.

3.2.2 *Main Takeaways*

3.2.2.1 *Main Points*

- A random variable is a real-valued function on the sample space that is of interest and it inherits a probability distribution from the probability measure.
- The probability distribution associated with a random variable encapsulates all relevant information about the random variable.
- The basic properties of common distributions such as the uniform distribution, binomial distribution, Poisson distribution, and normal distribution are reviewed.

3.2.2.2 *Main Terms*

- **Probability Density Function (PDF):** A statistical expression that defines a probability distribution (the likelihood of an outcome) for a discrete random variable (e.g., a stock or ETF) as opposed to a continuous random variable.

- **Poisson Distribution:** In statistics, a Poisson distribution is a statistical distribution that shows how many times an event is likely to occur within a specified period. It is used for independent events that occur at a constant rate within a given interval of time.
- **Binomial:** In probability theory and statistics, the binomial distribution with parameters n and p is the discrete probability distribution of the number of successes in a sequence of n independent experiments, each asking a yes–no question, and each with its Boolean-valued outcome: success/yes/true/one.

3.2.3 Definitions and Notations

We first state key definitions and introduce notations used for our discussion on probability distributions.

3.2.3.1 Sample Space

The *sample space* Ω is the set of all possible outcomes of an experiment or observation process.

3.2.3.2 Events

Events are subsets of the sample space. The sample space Ω is itself an event and it is a certain event as it is bound to happen since every observation is inside it. On the other hand, the empty set ϕ is also an event and it is the impossible event as it cannot happen. A *probability distribution* is a function P that assigns a real number between 0 and 1 inclusive to each event. For an event E, the number $P(E) \in [0, 1]$ is the probability of E happening. The closer $P(E)$ is to 1, the more likely it is to occur. In particular, $P(\Omega) = 1$ and $P(\phi) = 0$.

3.2.3.3 Random Variables

A *random variable* X is a function on the sample space Ω into the set of real numbers, R. We write $X : \Omega \to R$.

Example A: Toss a coin: sample space, $\Omega = \{\text{head}, \text{tail}\}$.

Consider the random variables X and Y defined below:
$$X(\omega) = \begin{cases} 1 & \text{if } \omega = \text{head} \\ 0 & \text{if } \omega = \text{tail} \end{cases}$$

$$Y(\omega) = \begin{cases} 1 & \text{if } \omega = \text{head} \\ -1 & \text{if } \omega = \text{tail} \end{cases}$$

If $X_1, X_2, \ldots, X_n$ are the results of n observations from tossing the coin, we notice that the sum $X_1 + X_2 + \cdots + X_n$ tells us the number of heads obtained in n tosses of the coin. What information can one get from $Y_1 + Y_2 + \cdots + Y_n$?

Example B: Roll two dice, say a red dice and a blue dice: sample space, $\Omega = \{(x, y) \mid x, y = 1, 2, 3, 4, 5, 6\}$.

Let us define random variables U, V and W as follows:

$$U(x, y) = x + y$$
$$V(x, y) = |x - y|$$
$$W(x, y) = x - 2y^2$$

Exercise: Describe the events $U = 7$, $V = 3$, $V = 6$, $W = -2$, and $U = 2V$.

Example C: Roll two dice say a red dice and a blue dice: sample space, $\Omega = \{(x, y) \mid x, y = 1, 2, 3, 4, 5, 6\}$.

Suppose the red and blue dice are both fair, that is each of the outcomes 1, 2, 3, 4, 5, 6 is equally likely, so $P((1,1)) = P((1,2)) = \cdots = P((6,6)) = 1/36$.

Exercise: What are the probabilities of the following events?

$$U = 7, \quad V = 3, \quad V = 6, \quad W = -2, \quad \text{and} \quad U = 2V?$$

That is, compute the probabilities $P(U = 7)$, $P(V = 3)$, $P(V = 6)$, $P(W = -2)$, $P(U = 2V)$.

3.2.3.4 *Discrete Random Variables*

If a random variable X takes on a countable number of values, $x_1, x_2, \ldots, x_n, \ldots$, we say that X is a discrete random variable. The following are important characteristics of a discrete probability distribution:

- The probability of a particular outcome is between 0 and 1 inclusive. That is, for all outcomes $x_1, x_2, \ldots x_n, \ldots$, $0 \leq P(X = x_n) \leq 1$.
- The outcomes are mutually exclusive events. That is, $\{X = x_i\} \cap \{X = x_j\} = \emptyset$ for $i \neq j$.
- The list of outcomes is exhaustive. That is, $P(X = x) = 0$ for all $x \notin \{x_1, x_2, \ldots, x_n, \ldots\}$. So the sum of the probabilities of the events $\{X = x_i\}$ for $i = 1, 2, 3, \ldots$, is equal to 1. That is,

$$P(X = x_1) + P(X = x_2) + \cdots + P(X = x_n) + \cdots$$
$$= \sum_{i \geq 1} P(X = x_i) = 1.$$

The function, $f(x_n) = P(X = x_n)$ for $n = 1, 2, 3, \ldots$, is called the probability mass function.

Note: upper case X is a random variable and lower case x_i is a value (real number) that the random variable takes on.

Let X be a discrete random variable taking on values $x_1, x_2, x_3, \ldots, x_n, \ldots$. Suppose

$$P(X = x_i) = p_i \quad \text{for } i = 1, 2, 3, \ldots.$$

The *expected value* of X or *mean of the distribution* is given by:

$$E(X) = x_1 p_1 + x_2 p_2 + \cdots + x_n p_n + \cdots = \sum_{i=1}^{\infty} x_i p_i.$$

The expected value of X or the mean value of the distribution is usually denoted by $\mu = E(X)$.

The *variance* of X measures how widely dispersed a typical observation is from the mean $E(X) = \mu$. It is defined as:

$$\mathrm{Var}(X) = E[(X - E(X))^2] = E(X^2) - E(X)^2$$
$$= \sum_{i=1}^{\infty} x_i^2 p_i + \left(\sum_{i=1}^{\infty} x_i p_i \right)^2 = E(X^2) - \mu^2.$$

It is usually denoted by $\sigma^2 = \mathrm{Var}(X)$.

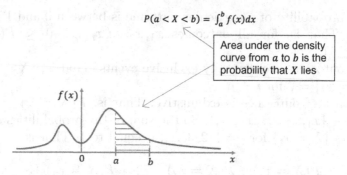

Figure 3.1: Probability distribution of a continuous random variable.

3.2.3.5 *Continuous Random Variables*

If a random variable X takes on an uncountable number of values, X is called a continuous random variable. To specify the probability that X takes on a value within a range, we use a PDF f.

The probability that X takes on a value between a and b is denoted by Figure 3.1.

Let X be a continuous random variable with density function $f(x)$. The expected value of X or mean of the distribution is given by

$$E(X) = \mu = \int_{-\infty}^{\infty} x \cdot f(x)\, \mathrm{d}x.$$

And the variance of X,

$$\mathrm{Var}(X) = \sigma^2 = E[(X - E(X))^2] = E(X^2) - E(X)^2$$
$$= \int_{-\infty}^{\infty} x^2 \cdot f(x)\, \mathrm{d}x - \left(\int_{-\infty}^{\infty} x \cdot f(x)\, \mathrm{d}x \right)^2.$$

3.2.3.6 *Sum of Random Variables: Mean and Variance of the Sum*

Let X and Y be random variables. Consider the sum $S = X + Y$.

The mean, $E(S) = E(X + Y) = E(X) + E(Y)$.

The covariance of X and Y is defined as $\mathrm{Cov}(X, Y) = E(XY) - E(X)E(Y)$.

If X and Y are independent, then $\mathrm{Cov}(X, Y) = 0$.

The variance of S,

$$\begin{aligned}
\text{Var}(S) &= E(S^2) - (E(S))^2 \\
&= E(X^2 + 2XY + Y^2) - (E(X) + E(Y))^2 \\
&= E(X^2) + 2E(XY) + E(Y^2) - (E(X)^2 \\
&\quad + 2E(X)E(Y) + E(Y)^2) \\
&= E(X^2) - E(X)^2 + E(Y^2) - E(Y)^2 \\
&\quad + 2[E(XY) - E(X)E(Y)] \\
&= \text{Var}(X) + \text{Var}(Y) + 2\text{Cov}(X, Y)
\end{aligned}$$

If X and Y are independent, then $\text{Var}(S) = \text{Var}(X+Y) = \text{Var}(X) + \text{Var}(Y)$.

The results above may be generalized to any linear combination of random variables.

Let $Y_1, Y_2, \ldots, Y_n$ be random variables with mean $E(Y_i) = \mu_i$ and variance $\text{Var}(Y_i) = \sigma_i^2$ for $i = 1, 2, \ldots, n$.

For constants $a_1, a_2, \ldots, a_n$ and b, the expected value or mean of the weighted sum:

$S = a_1 Y_1 + a_2 Y_2 + \cdots + a_n Y_n + b$ is given by:

$$E(S) = E(a_1 Y_1 + a_2 Y_2 + \cdots + a_n Y_n + b) = a_1 \mu_1 + a_2 \mu_2 + \cdots + a_n \mu_n + b.$$

Further if $Y_1, Y_2, \ldots, Y_n$ are independent or uncorrelated then the variance of S is given by:

$$\text{Var}(S) = \text{Var}(a_1 Y_1 + a_2 Y_2 + \cdots + a_n Y_n + b) = a_1^2 \sigma_1^2 + a_2^2 \sigma_2^2 + \cdots + a_n^2 \sigma_n^2.$$

3.3 Discrete Probability Distributions

Having provided the key definitions and notations, we now review the uniform, binomial, Poisson, and normal distributions and their properties.

A random variable X with a discrete probability distribution takes on only countably many possible outcomes $x_1, x_2, \ldots, x_n, \ldots$.

Common discrete probability distributions include the binomial and Poisson distributions as well as the discrete uniform distribution, which counts for one of the simplest forms of a discrete distribution.

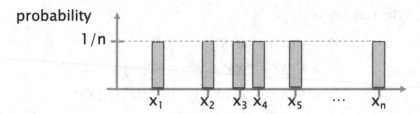

Figure 3.2: Probability mass function for a uniform discrete distribution.

The discrete uniform distribution for the outcomes of throwing a single fair dice can be expressed by the equations: $P(X = x) = 1/6$, where $x = 1, 2, 3, 4, 5$ or 6.

The mean of the discrete uniform distribution is:

$$\mu = \sum_{x=1}^{6} x \cdot P(X = x) = 1 \times \frac{1}{6} + 2 \times \frac{1}{6} + 3 \times \frac{1}{6}$$

$$+ 4 \times \frac{1}{6} + 5 \times \frac{1}{6} + 6 \times \frac{1}{6} = 7/2 = 3.50.$$

The variance of this discrete uniform distribution can be calculated using

$$\sigma^2 = \sum_{x=1}^{6} (x - \mu)^2 P(X = x) = \left(1 - \frac{7}{2}\right)^2 \cdot \frac{1}{6} + \left(2 - \frac{7}{2}\right)^2 \cdot \frac{1}{6}$$

$$+ \cdots + \left(6 - \frac{7}{2}\right)^2 \cdot \frac{1}{6} = \frac{35}{12} = 2.92.$$

The probability mass function for a discrete uniform distribution is shown in Figure 3.2.

3.3.1 *Binomial Distribution*

Binomial distribution is one of the most widely used discrete probability distributions. It is used in the binomial model for pricing financial derivatives.

The following are the characteristics of a binomial distribution.

- The outcome of an experiment is classified into one of two mutually exclusive categories: a success or a failure. We repeat the experiment n times and the outcome of each trial or experiment is independent of the outcomes of the other trials. The number of successes, X, among the n trials is a random variable that takes on the values $0, 1, 2, 3, \ldots, n$. If p denotes the probability that the outcome of an experiment is a success, then X has the binomial distribution with parameters, n and p and we write $X \sim B(n, p)$.

Let us consider an example. Each student in a large institution is either enrolled in the School of Business or not. It is known that $p = 5\%$ of the students are enrolled in the School of Business. Suppose we pick a random sample of five students. Let X denote the number of students in the sample who are enrolled in the School of Business. The random variable X takes on the values $0, 1, 2, 3, 4$, or 5 and it has the binomial distribution $B(5, 0.05)$. Strictly speaking, $X \sim B(5, 0.05)$ only when we sample with replacement. However, when the number of students in the institution is very large, the distribution of X is practically the binomial distribution $B(5, 0.05)$.

To specify a binomial distribution, we need two parameters, namely the number of trials n and the probability of a success p. If $X \sim B(n, p)$, the probability that $X = x$ where $x \in \{0, 1, 2, \ldots, n\}$ is given by

$$P(X = x) = \binom{n}{x} p^x (1 - p)^{n-x}$$

where $\binom{n}{x} = \frac{n!}{x!(n-x)!}$ Denotes the number of ways of selecting x objects from n distinct objects,

n is the number of independent trials,

p is the probability of success in a trial.

The mean μ and variance σ^2 of the binomial distribution $B(n, p)$ are given by

Mean of the binomial distribution: $\mu = np$

The variance of the binomial distribution: $\sigma^2 = np(1 - p)$.

Figure 3.3 depicts the probability mass functions for the binomial distributions $B(10, 0.5)$ and $B(20, 0.7)$ in the left and right panels, respectively.

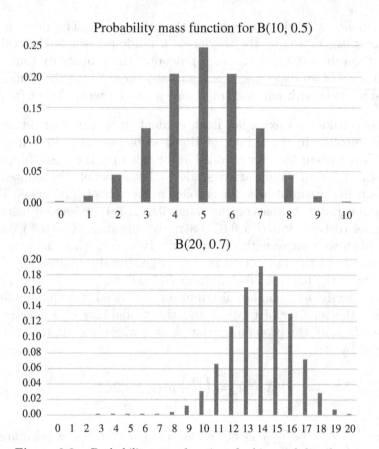

Figure 3.3:   Probability mass functions for binomial distributions.

Example

Suppose on a trading day, the share price of Singapore Airlines goes up with probability $p = 0.55$ independent of the moves on previous days. If A denotes the event that the share price has a run of 5 consecutive trading days of upward movement, what is the probability that A occurs?

$$P(A) = 0.55 \times 0.55 \times 0.55 \times 0.55 \times 0.55 = (0.55)^5 = 0.0503$$

Let X denote the number of days over a five-day period where the stock went up. Then $X \sim B(5, 0.55)$. The expected or average value of X and its variance are given by

Expected value or mean of X, $\quad E(X) = np = 5 \times 0.55 = 2.750$

The variance of X, $\text{Var}(X) = np(1 - p) = 1.2375$

3.3.2 *Poisson Distribution*

The Poisson distribution is another example of a discrete distribution. It describes the number of times some event occurs within a specified period, area, distance interval, or volume. For example, it can be used to model the number of new cars sold by a salesperson in a week or the number of dining customers queuing at a popular dumpling restaurant during lunch hour. The Poisson distribution is completely specified by one parameter λ, which turns out to be the mean of the distribution.

The following are the characteristics of a Poisson distribution (λ):

- The random variable X denotes the number of times an event occurs and thus X takes on values $0, 1, 2, 3, \ldots$. The probability that $X = x$ events occurred is given by:

$$P(X = x) = \frac{e^{-\lambda}\lambda^x}{x!}$$

where λ is a constant parameter and e is the natural number 2.71828.
- The mean μ and variance σ^2 of the Poisson distribution Poisson(λ) are given by

Mean of the Poisson distribution, $\mu = \lambda$

Variance of the Poisson distribution, $\sigma^2 = \lambda$.

Useful fact: if $X \sim \text{Poisson}(\lambda)$ and $Y \sim \text{Poisson}(\eta)$ are independent, then their sum $X + Y$ is a random variable whose distribution is the Poisson distribution, Poisson($\lambda + \eta$). So the sum of independent Poisson random variables is again a Poisson random variable. Let $X_i \sim \text{Poisson}(\lambda_i)$ for $i = 1, 2, \ldots, n$ be n independent Poisson random variables. Then their sum

$$X_1 + X_2 + \cdots + X_n \sim \text{Poisson}(\lambda_1 + \lambda_2 + \cdots + \lambda_n).$$

Example

Some analysts suggest that the number of days in a year that the US stock market loses 5% or more in a day may be modeled as a Poisson distribution. Is this reasonable? Losing 5% or more in a day is a rare event. For a day in a typical year, this represents roughly a 4 to 5 standard deviation move. For the first half of 2020 (1H 2020), there were 6 days where the daily return of the S&P500 Index was -5% or lower. From the year-to-date (YTD) data, January to mid-August 2020, there were still 6 such days. We may be interested to estimate the number of such days in 2020. How can we go about it?

The first step is to understand the phenomenon. Looking at historical information is a start and we need to keep in mind that future behavior can be very different from the past especially if the probability distribution is not stationary, that is, the distribution changes over time or with changes in various factors.

Figure 3.4 illustrates the number of days in a year that the S&P500 Index lost $R\%$ or more in a day for $R = 0, 1, 2, 3, 4, 5$, in a day, from 2007 till the first half of 2020. We see that in many years, the number of days that the index lost 5% or more is zero. In 2008 and 2009, there were 11 and 2 days respectively, in response to the US subprime mortgage and credit crisis. In the first half of 2020, the six observations were due to the COVID-19 pandemic. Are there reasonable assumptions that we can make to help us estimate the number of such days for the second half of the year? Is the loss on one-day independent of the returns on the other days?

S&P 500 Index data

	2007	2008	2009	2010	2011	2012	2013	2014	2015	2016	2017	2018	2019	1H 2020
Number of days in the year with return ≤ -5%	0	11	2	0	1	0	0	0	0	0	0	0	0	6
Number of days in the year with return ≤ -4%	0	15	6	0	4	0	0	0	1	0	0	1	0	9
Number of days in the year with return ≤ -3%	2	24	12	5	6	0	0	0	3	1	0	5	1	14
Number of days in the year with return ≤ -2%	12	41	28	10	21	3	2	5	6	5	0	15	5	21
Number of days in the year with return ≤ -1%	35	75	56	37	48	21	17	19	31	22	4	32	15	31
Number of days in the year with return ≤ 0%	114	127	112	108	114	118	105	108	133	122	108	119	102	56

Figure 3.4: S&P 500 index data set.

The use of the Poisson distribution to model the number of days in a year that the index falls 5% or more is thus problematic. The number of such days is driven largely by shocks and we need to understand the nature of such shocks to get a handle on this variable.

Of the six observations in the first half of 2020, five of them happened in March and one in June. If the COVID-19 situation improves or if an effective treatment or vaccine is found, then there may be zero such days in the second half of 2020.

On the other hand, if the situation worsens we may see many such occurrences. Thus the estimation of the number of such occurrences requires more in-depth analysis than can be covered in an introductory statistics course. Knowing the limitations of our tools is important for otherwise, we fool ourselves into thinking we have a solution to a problem when we don't.

3.3.2.1 *Poisson Distribution and Binomial Distribution*

The Poisson distribution can be used to approximate a binomial distribution $B(np)$ when the probability p of success on a single trial is very small and n is large. What parameter λ should we use for the Poisson distribution Poisson(λ)? If we wish to approximate $B(n, p)$ using Poisson(λ), it is natural that their means should agree. That is, we set:

Mean of Poisson (λ) distribution

$$= \lambda = np = \text{mean of } B(np) \text{ distribution.}$$

Since we have a closed-form formula for the probabilities of the binomial distribution, namely

$$P(X = x) = \binom{n}{x} p^x (1 - p)^{n-x}$$

Why bother with approximations? The problem is that $\binom{n}{x}$ is very difficult to compute for large n and moderate x.

3.4 Continuous Probability Distributions

A random variable with a continuous probability distribution assumes an uncountable number of values. As an example, the weights for a sample of small engine blocks are 54.3, 52.7, 53.1, and 53.9 kg. Here more precision may be provided if necessary. For example, the first value could be 54.3283 kg. For practical purposes, the weight may take on any real number greater than 0.

The two common families of continuous probability distributions are the uniform distribution and the normal distribution.

3.4.1 *Uniform Distribution*

The uniform distribution is the simplest distribution for a continuous random variable. The following are the characteristics of a uniform distribution:

- The random variable X takes values in a range $[a, b]$ where $a < b$ are finite real numbers. The endpoints of the interval may or may not be in the range of X.
- Each value in the interval (a, b) has an equal chance of occurring. The density curve over the interval (a, b) is a horizontal straight line. Outside the interval $[a, b]$, the density curve takes on the value 0.

The probability density curve for the uniform distribution is given by

$$f(x) = \begin{cases} \dfrac{1}{b-a} & a \leq x \leq b \\ 0 & \text{otherwise} \end{cases}$$

A graphical representation of the uniform probability distribution $U(a, b)$ is shown in Figure 3.5.

$$\text{Mean of the Uniform Distribution, } \mu = \frac{a+b}{2}$$

$$\text{The variance of the Uniform Distribution, } \sigma^2 = \frac{(b-a)^2}{12}$$

3.4.2 *Normal Distribution*

The normal distribution is the most important continuous probability distribution and it is uniquely determined by two parameters:

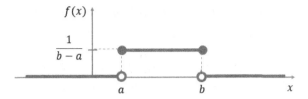

Figure 3.5: Graphical representation of uniform probability distribution.

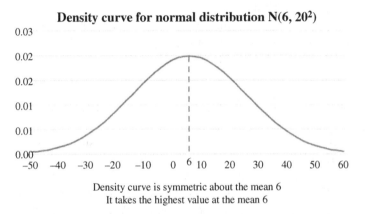

Density curve is symmetric about the mean 6
It takes the highest value at the mean 6

Figure 3.6: Illustrated density curve for normal distribution $N(6, 20^2)$.

the mean μ and standard deviation σ. A random variable X has the normal distribution $N(\mu, \sigma^2)$ if its density function is given by

$$f(x) = \frac{1}{\sigma\sqrt{2\pi}} e^{-\frac{1}{2}\left(\frac{x-\mu}{\sigma}\right)^2} \quad \text{for } -\infty < x < \infty.$$

Its shape is the famous bell curve. The density curve for the normal distribution $N(6, 20^2)$ is depicted in Figure 3.6.

The major characteristics of the normal distribution are:

- The density function for a normal distribution is "bell-shaped" and the mean, median, and mode are all equal and are located at the center of the distribution.
- The density function is symmetrical about the mean.
- The density function $f(x)$ approaches zero as x goes to $\pm\infty$.

Useful fact: if X and Y are independent normal random variables, then for any real numbers a and b (other than $a = b = 0$), the sum $S = aX + bY$ is again a normal random variable.

3.4.3   *Standard Normal Probability Distribution*

The standard normal distribution is $N(0,1)$.

Every normal random variable $X \sim N(\mu, \sigma^2)$ is related to the standard normal random variable through the relationship

$$Z = \frac{X - \mu}{\sigma}.$$

It is easy to verify that $Z \sim N(0,1)$.

Conversely for any real numbers a and b where $a \neq 0$, the random variable $Y = aZ + b$ has the normal distribution $N(b, a^2)$.

A useful rule of thumb for $X \sim N(\mu, \sigma^2)$.

- About 68% of the area under the normal density curve is within plus one and minus one standard deviation of the mean. That is, $P(\mu - \sigma < X < \mu + \sigma) = 68\%$.
- About 95% of the area under the normal density curve is within plus and minus two standard deviations of the mean. That is, $P(\mu - 2\sigma < X < \mu + 2\sigma) = 95\%$.
- Practically all of the area under the normal density curve is within three standard deviations of the mean. We note that, $P(\mu - 3\sigma < X < \mu + 3\sigma) = 99.73\% \cong 100\%$.

Figure 3.7 illustrates the rules for the normal distribution $N(6, 20^2)$.

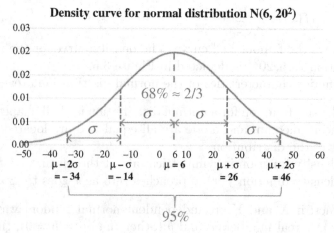

Figure 3.7:   Rules illustrated for density curve of normal distribution $N(6, 20^2)$.

Bibliography

Illowsky, B. and Dean, S. (2013). *Introductory Statistics.* OpenStax. Retrieved from openstar.org.

Triola, M. (2018). *Elementary Statistics.* 13th ed., Chaps. 4–6. Pearson, New Jersey, NJ.

3.5   Sample Questions

Please select the most appropriate response.

The following information pertains to Questions 1 to 2.

Let X and Y denote the outcomes of two independent rolls of a fair dice.

Question 1

What is the probability that $|X - Y| > 4$?

(a) $1/6 = 0.167$
(b) $1/9 = 0.111$
(c) $1/18 = 0.056$

Question 2

What is $P(2X = 3Y)$?

(a) $1/36 = 0.028$
(b) $1/18 = 0.056$
(c) $1/9 = 0.111$

The following information applies to Questions 3 to 6.

Let X be a random variable whose density function is given by

$$f(x) = \begin{cases} 2x, & 0 \le x \le 1 \\ 0, & \text{otherwise} \end{cases}.$$

Question 3

What is the mean value or expected value of X?

(a) $E(X) = 1/2$
(b) $E(X) = 2/3$
(c) $E(X) = 5/6$

Question 4

What is the standard deviation of the distribution?

(a) 0.2357
(b) 0.5000
(c) 0.5735

Question 5

Compute the probability that $X < 0.7$.

(a) 0.70
(b) 0.63
(c) None of the above.

Question 6

Compute the expected value of $X^2 - 5$.

(a) $E(X^2 - 5) = -2.5$
(b) $E(X^2 - 5) = -4$
(c) $E(X^2 - 5) = -4.5$

The following information applies to Questions 7 to 9.

Let Z_1 and Z_2 be two independent random variables. Suppose they both have the standard normal distribution, $N(0,1)$.

Question 7

What is the expected value of $2Z_1 - Z_2$?

(a) -1
(b) 0
(c) 2

Question 8

What is the variance of $2Z_1 - Z_2$?

(a) 1
(b) 3
(c) None of the above.

Question 9

What is the probability that $2Z_1 - Z_2$ lies in the interval (0,3)? That is, compute the probability $P(0 < 2Z_1 - Z_2 < 3)$?

(a) 0.41
(b) 0.32
(c) 0.25

The following information applies to questions Questions 10 to 12.

A large school claimed that 80% of its students do volunteer work in charitable organizations. A newly hired teacher did a random survey with 7 students and found that only 1 had volunteered in some charities.

Question 10

Assuming that the school's claim of 80% is valid, what is the probability of obtaining the result in the survey?

(a) 0.000013
(b) 0.000358
(c) 0.002097

Question 11

Assuming that the school's claim is valid, what is the probability that none of the students in the sample of 7 students carried out volunteer work?

(a) 0.000013
(b) 0.000358
(c) 0.002097

Question 12

What may be a reasonable conclusion that can draw from the survey?

(a) The sample size was too small for us to make any inference, it must be 30 or larger.
(b) The result of the survey was probably consistent with the claim as atypical samples may be obtained by random chance.
(c) If the claim is valid, then the chance of having 1 or fewer students in the survey who did volunteer work was less than 0.04%. This extremely rare event suggested that the claim of 80% was probably too high.

Solutions

Question 1

Solution: Option **c** is correct

The event $|X - Y| > 4$ comprises the outcomes $(X, Y) = (1, 6), (6, 1)$. Thus $P(|X - Y| > 4) = 2/36 = 1/18$.

Question 2

Solution: Option **b** is correct

There are only two cases out of the 36 possible outcomes that result in $2X = 3Y$, namely $(X, Y) = (3, 2)$ and $(6, 4)$.

Question 3

Solution: Option **b** is correct.

The mean $E(X) = \int_0^1 x f(x) dx = \int_0^1 2x^2 dx = \left[\frac{2}{3} x^3\right]_0^1 = \frac{2}{3}$.

Question 4

Solution: Option **a** is correct

We first compute $E(X^2) = \int_0^1 x^2 f(x) dx = \int_0^1 2x^3 dx = \left[\frac{1}{2} x^4\right]_0^1 = \frac{1}{2}$.

The variance $\sigma^2 = E(X^2) - (E(X))^2 = \frac{1}{2} - \left(\frac{2}{3}\right)^2 = \frac{1}{18}$.

The standard deviation $\sigma = \frac{1}{\sqrt{18}} = 0.2357$.

Question 5

Solution: Option **c** is correct

The required probability $P(X < 0.7) = \int_0^{0.7} 2x dx = [x^2]_0^{0.7} = 0.49$.

Question 6

Solution: Option **c** is correct

We computed $E(X^2) = \frac{1}{2}$ earlier. Thus $E(X^2 - 5) = E(X^2) - 5 = -4.5$.

Question 7

Solution: Option **b** is correct

$$E(2Z_1 - Z_2) = E(2Z_1) - E(Z_2) = 2E(Z_1) - E(Z_2) = 2 \times 0 - 0 = 0$$

Question 8

Solution: Option **c** is correct

$$\mathrm{Var}(2Z_1 - Z_2) = 2^2\mathrm{Var}(Z_1) + (-1)^2\mathrm{Var}(Z_2) = 4 + 1 = 5.$$

Question 9

Solution: Option **a** is correct
We note that $Y = 2Z_1 - Z_2 \sim N(0.5)$. The required probability

$$P(0 < Y < 3) = P\left(0 < Z < \frac{3}{\sqrt{5}}\right) = 0.41.$$

Question 10

Solution: Option **b** is correct
Let X denote the number of students in the survey who did volunteer work. Then $X \sim B(7, 0.8)$. The probability that only 1 student did volunteer work, $P(X = 1) = \binom{7}{1} \times (0.8) \times (0.2)^6 = 0.000358$.

Question 11

Solution: Option **a** is correct
The probability that none of the students did volunteer work,

$$P(X = 0) = \binom{7}{0} \times (0.8)^0 \times (0.2)^7 = 0.000013.$$

Question 12

Solution: Option **c** is correct
The probability of obtaining the result of something more extreme against the claim was $P(X \leq 1) = P(X = 0) + P(X = 1) = 0.000371 < 0.04\%$. This extremely low probability suggested that the claim of 80% was likely to be invalid.

Chapter 4

Sampling and Estimation

4.1 Sampling

Sampling is a process used in statistical analysis in which a predetermined number of observations are taken from a larger population.

4.1.1 *Learning Objectives*

- Discuss how useful information about a population is obtained from a sample.

4.1.2 *Main Takeaways*

4.1.2.1 *Main Points*

- We sample from a population for a variety of reasons and need to interpret information obtained from the sample about the population.
- Central limit theorem states that for any distribution with finite variance, the sampling distribution of the sample mean of a random sample drawn from the distribution when properly normalized approaches the standard normal distribution.

4.1.2.2 *Main Terms*

- **Sampling Error:** A statistical error that occurs when an analyst does not select a sample that represents the entire population of

data and the results found in the sample do not represent the
results that would be obtained from the entire population.

- **Sampling Distribution of Sample Mean** can be thought of
  as "For a sample of size n, the sample mean will behave according to this distribution." Any random draw from that sampling
  distribution would be interpreted as the mean of a sample of n
  observations from the original population.
- **Standard Error of the Sample Mean:** Put simply, the standard error of the sample mean is an estimate of how far the sample
  mean is likely to be from the population mean, whereas the standard deviation of the sample is the degree to which individuals
  within the sample differ from the sample mean.

4.1.3　*Why Sample*

We study samples drawn from a population to obtain useful information about the population characteristics such as its mean and
variance.

A population refers to the entire group of objects or persons of
interest. The population of interest might be all the persons in rural
districts without a bank account or all telecommunication users still
using landlines.

A sample is a portion, a part, or a subset of the population. We
study samples to understand certain features of the population. The
observations in a sample are usually chosen so that they are independent of each other. Also, in some situations where we have no
or little information about the original distribution from which we
took our sample, we are still able to make reasonable statements
about the distribution of the sample means using the central limit
theorem.

There are several reasons why sampling is necessary:

- To survey or examine the entire population is time-consuming.
- The cost of studying all the items in the population is often
  prohibitive.
- The destructive nature of certain tests. If we were to test every
  can of beans in a shipment to determine whether they are fit for
  consumption, we have nothing left to sell.

4.1.4 *Sampling Error*

It is unlikely that the results obtained from a sample will coincide exactly with those from a population. For example, it is unlikely that the minute usage per user of the sample of 50 landline users is the same as that of the population. Sometimes, the presence of outliers in the sample can result in significant sampling errors, for example, the difference between a sample statistic and its corresponding population parameter.

4.1.5 *Sampling Distribution of the Sample Mean*

Let $X_1, X_2, \cdots, X_n$ be the observations obtained from a random sample of size n. The average of these observations is the sample mean is:

$$\bar{X} = \frac{X_1 + X_2 + \cdots + X_n}{n}.$$

Suppose all possible samples of the size n are selected from a specified population, and the mean of each of these samples is computed. The distribution of these sample means is called the sampling distribution of the sample mean $\bar{X}$. Formally, a sampling distribution of the sample mean can be defined as a probability distribution of all possible sample means of a given sample size.

The sampling distribution of the sample mean has the following characteristics:

- The mean of all the sample means will be exactly equal to the population mean.
- If the population from which the samples are drawn is normally distributed, the sample mean will be normally distributed.
- If the population from which the samples are drawn is not normally distributed but has a finite variance, the sampling distribution of the sample mean is approximately normal, provided the samples are sufficiently large.

4.1.6   *The Central Limit Theorem*

The central limit theorem states that if all samples of a specified size are selected from any population with finite variance, the sampling distribution of the sample means is approximately a normal distribution for large sample sizes. The approximation improves with larger samples. This means that for large random samples, the shape of the sampling distribution of the sample means is close to a normal probability distribution when properly scaled. Using the central limit theorem, we can make reasonable statements about the distribution of sample means with little or no information about the shape of the original distribution from which we took the sample.

4.1.7   *Standard Error of the Sample Mean*

Let the population standard deviation be denoted by σ. It is easy to show that the standard deviation of the sampling distribution of the sample mean is $\frac{\sigma}{\sqrt{n}}$, where n is the number of observations in each sample. The assumption that the observations in a sample are uncorrelated is needed here. As the observations are independent, they will be uncorrelated too. We refer to $\frac{\sigma}{\sqrt{n}}$ as the standard error of the sample mean.

$$\text{Standard Error of the Sample Mean, } \sigma_{\bar{X}} = \frac{\sigma}{\sqrt{n}}$$

where σ is the population standard deviation and n is the sample size.

The size of the standard error is affected by the standard deviation of the population and the sample size. As the standard deviation of the population increases so does the standard error of the sample mean. As the sample size increases, the standard error decreases, which indicates that there is less variability in the distribution of the sample means.

4.1.8   *Using the Sampling Distribution of the Sample Mean*

Many business decisions are made based on sampling. Generally, we have a population and we wish to know something about the population, such as the mean value of some quantity. For example, we may want to estimate the average disposable income per worker or the average cholesterol level of adults in a country. We take a sample from the population and make an inference about the population parameter using the data in the sample.

We distinguish between the following two cases:

Case 1: When a random sample of size n is taken from a population that follows the normal distribution $N(\mu, \sigma^2)$, the sampling distribution of the sample mean $\bar{X}$ will be the normal distribution, $N(\mu, \frac{\sigma^2}{n})$.

Case 2: Suppose the distribution for the population is not known, but we can assume that its variance σ^2 is finite, then the sampling distribution of the sample mean $\bar{X}$ maybe approximated by a normal distribution when the sample contains 30 or more observations.

We have $\bar{X} \sim N\left(\mu, \frac{\sigma^2}{n}\right)$ exactly in case 1 and approximately in case 2.

With the normalization, $= \frac{\bar{X}-\mu}{\sigma/\sqrt{n}}$ we see that $Z \sim N(0, 1)$ the standard normal distribution.

4.2 Estimation and Confidence Intervals

4.2.1 *Estimation*

Estimation of an unknown quantity is often a key objective of statistical analysis. Besides point estimates, confidence interval (CI) estimates give a range of values that contain the quantity of interest with high probability.

Using a sample, we form estimators for a population characteristic such as its mean. What are the desirable properties of an optimal estimator? Are unbiased estimators the only game in town? Often when several estimators are available, a criterion such as the least-square error may be used to pick the optimal estimators.

Estimation of parameters of distributions or models and other quantities is of practical importance. Suppose a marketing company would like to estimate the proportion of adults in Singapore who prefer Coca-Cola to Pepsi Cola. How can we go about estimating this? If we have the resources to obtain the preference of all adults in Singapore, then we would have an accurate answer. If not, we use a random sample that is representative of the adult population and from it, estimate the proportion.

CI refers to a range of values within which statisticians believe the actual value of a certain population parameter lies. It differs from a point estimate which is a single, specific numerical value. Unlike point estimates, a CI estimate requires some knowledge or assumption about the sampling distribution of the estimator.

4.2.1.1 *Learning Objectives*

- Discuss the CI estimate of a population characteristic.

4.2.1.2 *Main Takeaways*

4.2.1.2.1 Main Points

- Using a sample, we form estimators for a population characteristic.
- When constructing confidence intervals, we must specify the probability that the interval contains the true value of the parameter of interest. This probability is represented by $(1 - \alpha)$ where α is the level of significance. In statistical terminology, $1 - \alpha$ is called the degree of confidence or certainty.
- A random sample of size n is said to have an $n - 1$ degree of freedom for estimating the population variance, in the sense that there is only $n - 1$ independent deviation from the mean on which to base the estimate.
- The t-distribution has fatter tails than the standard normal distribution but converges to the standard normal distribution as degrees of freedom go to infinity.
- The desirable properties of an estimator are unbiasedness (the expected value of the estimator equals the population parameter), efficiency (the estimator has the smallest variance), and consistency (the probability of accurate estimates increases as sample size increases).

4.2.1.2.2 Main Terms

- **Point Estimate:** refers to a calculated value of the sample statistic such as the mean, X.
- **Symmetrical:** the correspondence of the form and arrangement of elements or parts on opposite sides of a dividing line or plane or about a center or an axis.

- **Mean Squared Error (MSE):** also known as mean squared deviation (MSD) of an estimator (of a procedure for estimating an unobserved quantity) measures the average of the squares of the errors—that is, the average squared difference between the estimated values and the actual value.
- **Unbiased Estimator:** a statistic is said to be an *unbiased estimate* of a given parameter when the mean of the sampling distribution of that statistic can be shown to be equal to the parameter being *estimated*.
- **Confidence interval (CI):** a range of values within which statisticians believe the actual value of a certain population parameter lies. It differs from a point estimate which is a single, specific numerical value. Unlike point estimates, a CI estimate requires some knowledge or assumption about the sampling distribution of the estimator.

4.2.2 *Point Estimate*

A point estimate is a statistic, computed from the observations in a sample, which is used to estimate a population parameter. Below are some examples of point estimates:

- The sample mean, $\bar{X}$, is a point estimate of the population mean, μ.
- The sample standard deviation, s, is a point estimate of the population standard deviation, σ.

Example
Suppose we survey 1,000 adults chosen at random. For $i = 1, 2, \ldots, 1,000$, let the random variable X_i be defined as:

$$X_i = \begin{cases} 1 & \text{if the } i\text{th person prefers Coke} \\ 0 & \text{otherwise} \end{cases}$$

If we sample with replacement, the sum $S = X_1 + X_2 + \cdots + X_{1,000}$ has the binomial distribution $B(1,000, p)$ where p is the proportion (or probability) of adults preferring Coke. If we do not sample with replacement, the binomial distribution is a good approximation given that the adult population in Singapore is many times larger than 1,000.

The sample mean

$$\hat{p} = \frac{X_1 + X_2 + \cdots + X_{1,000}}{1,000} = \frac{S}{1,000}$$

is an unbiased estimate of p.

Suppose 529 people in the survey preferred Coke, then our estimate for p is

$$\hat{p} = \frac{S}{1,000} = 0.529 = 52.9\%.$$

This is a point estimate. How good this estimate depends on the criteria for assessing an estimator. Under the least squared error criteria, this is the best.

We don't believe that our point estimate of 52.9% is exactly equal to the true proportion p of people who prefer Coke. Could the true proportion be 55% or even 49%? Instead of using a number to estimate the true proportion p, we seek an interval that contains p with high probability.

4.2.3   Confidence Intervals

a. A CI is a range of values, constructed from the sample data so that the population parameter is likely to occur within that range with a specified probability. This specified probability is called the level of confidence.
b. The level of confidence is a measure of the confidence we have that an interval estimate will include the population parameter.

We say that a $W\%$ CI is an interval such that the probability that it contains the true proportion p is at least $W\%$. We must note that a statement such as we are 99.9% confident that the parameter p lies between 0.5% and 98% is of little help to us in trying to learn about the population proportion, p. What we like to have is a very small interval that contains p with high confidence or probability.

a. The central limit theorem helps us to construct confidence intervals for the population mean μ.
b. A 95% CI means that the probability that the given interval includes the actual (unknown) population mean is (at least) 95%.

c. A 99% CI means that the probability that the given interval includes the actual (unknown) population mean is (at least) 99%.

Population standard deviation σ is unknown.

We will consider the case where the population standard deviation σ is unknown. In this case, the sample standard deviation, s is used as an estimate of σ in the construction of confidence intervals.

Suppose $X_1, X_2, \cdots, X_n$ are independent random observations from a distribution with mean μ and finite variance σ^2. We usually use the "sample variance" s^2 to estimate σ^2. It is defined as:

$$s^2 = \frac{\sum_{i=1}^{n}(X_i - \bar{X})^2}{n - 1}$$

What can we say about s^2? It is an unbiased estimate for σ^2, that is, $E(s^2) = \sigma^2$. We can say more if we make assumptions about the underlying distribution.

Suppose $X_1, X_2, \cdots, X_n$ are independent random observations taken from a normal distribution $N(\mu, \sigma^2)$. It is a fact that the sum of the squares of n independent standard normal variables follows a chi-square distribution χ_n^2 with n degrees of freedom. Thus

$$\sum_{i=1}^{n} \left(\frac{X_i - \mu}{\sigma}\right)^2 \sim \chi_n^2.$$

Notice that in the definition of s^2 we use the sample mean $\bar{X}$ in place of μ since we do not know μ. It turns out that

$$(n - 1)\frac{s^2}{\sigma^2} = \sum_{i=1}^{n} \left(\frac{X_i - \bar{X}}{\sigma}\right)^2 \sim \chi_{n-1}^2.$$

Here, the degree is $n - 1$ because we estimated μ using $\bar{X}$.

4.2.4 *Student t-distribution*

Now let $Z \sim N(0, 1)$ and $Y \sim \chi_k^2$ be independent random variables having the standard normal distribution and the chi-square distribution with k degrees of freedom respectively.

Then, the ratio $t = \frac{Z}{\sqrt{Y/k}} \sim t_k$ has the t-distribution with k degrees of freedom.

The t-distribution has the following properties:

- Like the standard normal distribution, the t-distribution with k degrees of freedom t_k is continuous, bell-shaped, and symmetrical about the vertical axis, but it has fatter tails compared to the standard normal distribution.
- Instead of there being just one t-distribution, there is a "family" of t-distributions characterized by their degrees of freedom k. All the t-distributions have a mean of 0. For $k > 2$, the standard deviation of t_k is $\sqrt{\frac{k}{k-2}}$. As $k \to \infty$, t_k approaches the standard normal distribution $N(0, 1)$.

Suppose $X_1, X_2, \cdots, X_n$ are independent random observations taken from a normal distribution $N(\mu, \sigma^2)$. As usual, we have the sample mean,

$$\bar{X} = \frac{X_1 + X_2 + \cdots + X_n}{n} \quad \text{and sample variance, } s^2 = \frac{\sum_{i=1}^{n}(X_i - \bar{X})^2}{n-1}.$$

Observe that $Z = \frac{\bar{X}-\mu}{\sigma/\sqrt{n}} \sim N(0, 1)$ and $Y = (n-1)\frac{s^2}{\sigma^2} \sim \chi_{n-1}^2$. Moreover, it can be shown that Z and Y are independent. Thus we have:

$$\frac{\bar{X} - \mu}{s/\sqrt{n}} = \frac{\frac{\bar{X}-\mu}{\sigma/\sqrt{n}}}{\sqrt{s^2/\sigma^2}} = \frac{Z}{\sqrt{Y/n-1}} \sim t_{n-1}. \qquad (4.1)$$

The line (4.1) above will help us construct CI estimates for μ.

The steps to develop a CI for the population mean μ with an unknown but finite population standard deviation σ. We consider two cases.

Case 1: The population has the normal distribution $N(\mu, \sigma^2)$.

1. Estimate the population standard deviation, σ with the sample standard deviation, s.
2. Use the t-distribution with $n-1$ degree of freedom.

For the t-distribution with $n-1$ degrees of freedom t_{n-1}, we let $t_{n-1}(\alpha)$ denote the α-quantile of the distribution where $0 < \alpha < 1$. That is if $T \sim t_{n-1}$ then $P(T \le t_{n-1}(\alpha)) = \alpha$.

From (4.1) we have

$$P\left(t_{n-1}(\alpha/2) < \frac{\bar{X} - \mu}{s/s\sqrt{n}} < t_{n-1}(1 - \alpha/2)\right) = 1 - \alpha.$$

Note that $t_{n-1}(\alpha/2) = -t_{n-1}(1 - \alpha/2)$ due to the symmetry of the t-distribution about the vertical axis. Rearranging the expression, we see that

$$P(\bar{X} - t_{n-1}(1-\alpha/2) \cdot s/\sqrt{n} < \mu < \bar{X} + t_{n-1}(1-\alpha/2) \cdot s/\sqrt{n}) = 1 - \alpha.$$

Thus a $1 - \alpha$ CI for the population mean μ is

$$(\bar{X} - t_{n-1}(1 - \alpha/2) \cdot s/\sqrt{n}, \bar{X} + t_{n-1}(1 - \alpha/2) \cdot s/\sqrt{n}).$$

The end-points of the interval are usually written as $\bar{X} \pm t_{n-1}(1 - \alpha/2)) \cdot s/\sqrt{n}$.

Case 2: The sample size is 30 or larger.

1. Estimate the population standard deviation, σ with the sample standard deviation, s.

2. Then $\bar{X} \sim N\left(\mu, \frac{s^2}{n}\right)$ approximately.

$A_{1-\alpha}$ CI for the population mean is:

$$(\bar{X} - Z_{1-\alpha/2} \cdot s/\sqrt{n}, \bar{X} + Z_{1-\alpha/2} \cdot s/\sqrt{n})$$

where $Z \sim N(0,1)$ and $P(Z < Z_u) = u$.

Example
A factory makes syrup that is packed into 10-liter bottles. The engineer reckoned that the volume of syrup (in liters) in a bottle may be modeled as a normal random variable having the distribution $N(10, \sigma^2)$. For quality control purposes, five bottles were sampled at random and the volumes of syrup (in liters) were: 7.38, 10.97, 9.99, 9.31, 9.74.

What is a 95% CI for the population mean μ? We note that the population variance σ^2 is unknown.

Solution: We first compute $\bar{X} = (7.38+10.97+9.99+9.31+9.74)/5 = 9.48$.

The sample variance is

$$s^2 = \frac{(7.38 - \bar{X})^2 + (10.97 - \bar{X})^2 + (9.99 - \bar{X})^2 + (9.31 - \bar{X})^2 + (9.74 - \bar{X})^2}{5 - 1} = 1.75.$$

Sample standard deviation, $s = 1.32$.

Hence, we have $\frac{\bar{X}-\mu}{s/\sqrt{n}} = \frac{9.48-\mu}{1.32/\sqrt{5}} = \frac{9.48-\mu}{0.59} \sim t_4$.

Now $t_4(0.025) = -2.7765$ and $t_4(0.975) = 2.7765$, and $9.48 - 0.59 \times t_4(0.975) = 7.84$ and $9.48 + 0.59 \times t_4(0.975) = 11.12$.

Hence a 95% CI for the population mean μ is $(7.84, 11.12)$.

Example (continued from the Coke vs. Pepsi example)
We first find the sampling distribution of the sample mean or proportion

$$\hat{p} = \bar{X} = \frac{X_1 + X_2 + \cdots + X_{1,000}}{1,000} = \frac{S}{1,000}$$

We note that X_i is a Bernoulli random variable with mean p and variance $\sigma^2 = p(1-p)$. As $S \sim B(1,000, p)$, we see that its mean $E(S) = 1,000\,p$ and variance $\text{var}(S) = \sigma^2 = 1{,}000\,p\,(1-p)$.

The sample mean or proportion $\hat{p}$ has mean p and variance

$$\tau^2 = \text{Var}(S)/1{,}000^2 = \frac{1{,}000\sigma^2}{1{,}000^2} = \frac{p(1-p)}{1{,}000}.$$

By the central limit theorem, $\hat{p} \sim N(p, \tau^2) = N\left(p, \frac{p(1-p)}{1{,}000}\right)$ approximately.

Recall that a normal variable has a 95% probability of lying within 1.96 standard deviations of its mean. That is, $P(p - 1.96\,\tau < \hat{p} < p + 1.96\,\tau) = 95\%$.

We can re-write the inequality as

LHS: $p - 1.96\,\tau < \hat{p}$ is the same as $p < \hat{p} + 1.96\,\tau$
RHS: $\hat{p} < p + 1.96\,\tau$ is the same as $\hat{p} - 1.96\,\tau < p$

Putting the two together, we have $P(\hat{p} - 1.96\,\tau < p < \hat{p} + 1.96\,\tau) = 95\%$.

Hence, a 95% CI for p is $(\hat{p} - 1.96\,\tau, \hat{p} + 1.96\,\tau)$.

But what is τ? Well, $\tau^2 = p(1-p)/1{,}000$ but p is what we don't know! We can still work out a useful conclusion. Note that $p(1-p) \le 1/2 \times (1 - 1/2) = 1/4$ for all real p. So $\tau^2 \le 1/4{,}000$ and $\tau \le 0.0158$. Thus $1.96\,\tau \le 0.0310$.

From the survey, the sample proportion $\hat{p} = 0.529$. A 95% CI of p is $(0.529 - 0.0310, 0.529 + 0.0310) = (49.80\%, 56.00\%)$. However, the bound on τ makes the interval larger, and thus the interval is likely to have greater than 95% confidence. The chance that the proportion of p is between 49.80% and 56.00% is at least 95%.

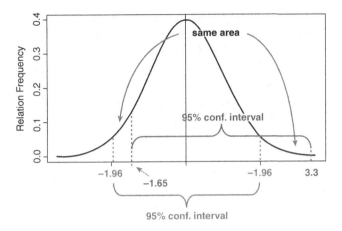

Figure 4.1: Illustration of 95% CI.

Remark: As we have a very large sample, we can obtain a sharper interval by estimating the population variance with the sample variance $s^2 = \left(\frac{n}{n-1}\right)\hat{p}(1-\hat{p})$. In practice, folks simply use $\hat{p}(1-\hat{p})$ as the difference is negligible when sample size n is large.

So $\tau^2 = \frac{s^2}{n} = \frac{\hat{p}(1-\hat{p})}{n-1} = 0.000249$ and $\tau = 0.015793 = 0.0158$ which is practically what we had earlier. The reason why the two numbers are so close is because $\hat{p} = 0.529$ is near 0.5.

So the earlier approximation is not too bad. In general, for large samples using s^2 as an estimate for the population variance gives a sharper result than bounding the population variance by 0.25.

By definition, there are many different 95% confidence intervals we can construct. Figure 4.1 is an example of the different 95% CI we can construct. For example,

$$P(p - 1.65\tau < \hat{p} < p + 3.30\tau) = 95\%.$$

From this, the 95% CI for p is given by Interval (A) $=$ $(50.29\%, 58.12\%)$. That is, there is a 95% chance that p lies between 50.29% and 58.12%. How does this compare with the 95% CI Interval (B) $= (49.80\%, 56.00\%)$ found earlier? Both intervals have the same confidence level of 95%. Which do you prefer?

Interval (A) has length $58.12\% - 50.29\% = 7.83\%$ while Interval (B) has length 6.20%. We usually prefer shorter intervals if they have the same confidence. Interval (B) was constructed from $P(\hat{p} - 1.96\,\tau < p < \hat{p} + 1.96\tau) = 95\%$ which is symmetric about $\hat{p}$. The shape of the normal distribution makes it the 95% CI with the shortest length.

There is a connection between interval estimates and hypothesis testing. Consider what the survey data tell us: with 529 people out of 1,000 surveyed preferring Coke, there is a 95% chance that the true population p lies between $(49.80\%, 56.00\%)$. In other words, at the 5% significance level (or $100\% - 5\% = 95\%$ confidence level), the data is consistent with any p in the interval $(49.80\%, 56.00\%)$, including $p = 50\%$. If $p = 53\%$ or if $p = 49\%$ say, we should not be surprised by the data. We are "surprised" if an event that has a 5% or less chance of happening, occurred, thus making it an unusual event. Since $p = 50\%$ is in the interval, we do not reject the null hypothesis $H_0 : p = 0.50$ in the two-tailed test at the 5% level of significance (Please see the section on "Hypothesis Testing").

4.2.5 *Unbiased Estimators*

Let $\hat{\theta}$ be an estimator for the parameter Θ. We say that $\hat{\theta}$ is an unbiased estimator if $E(\hat{\theta}) = \Theta$. That is, the average value of the estimator $\hat{\theta}$ will be the true value of the parameter Θ. The law of large numbers tells us that with an unbiased estimator there is a way to improve the estimate. Namely by making more estimates and taking their average value or by simply increasing the sample size.

Given an estimator $\hat{\theta}$ for the parameter Θ, we define the bias of the estimator as:

$$\text{Bias} = E(\hat{\theta}) - \Theta.$$

For an unbiased estimator, its bias is zero.

Example
Let $X_1, X_2, \cdots, X_n$ be independent random observations taken from a distribution with mean μ and finite variance σ^2.

The sample mean $\bar{X} = \frac{X_1 + X_2 + \cdots + X_n}{n}$ and sample variance $s^2 = \frac{\sum_{i=1}^{n}(X_i - \bar{X})^2}{n-1}$ are natural estimates for μ and σ^2, respectively. It is a good exercise to convince yourself that the two estimators are unbiased estimators. That is, $E(\bar{X}) = \mu$ and $E(s^2) = \sigma^2$.

4.2.6 *Criteria for Comparing Estimators*

Let $\hat{\theta}$ be an estimator for the parameter Θ. The error of the estimate is given by $\hat{\theta} - \Theta$ and $(\hat{\theta} - \Theta)^2$ is the square of the error.

The $\mathrm{MSE}(\hat{\theta}) = E(\hat{\theta} - \Theta)^2$. It is the average squared difference between the estimated values and the actual value. It is clear that the smaller the MSE, the smaller the error is on average.

It turns out that

$$\mathrm{MSE}(\hat{\theta}) = E(\hat{\theta} - \Theta)^2 = E(\hat{\theta} - E(\hat{\theta}))^2 + (E(\hat{\theta}) - \Theta)^2 = \mathrm{Var}(\hat{\theta}) + \mathrm{Bias}^2$$

For an unbiased estimator, its MSE is simply its variance.

Given two estimators $\hat{\theta}_1$ and $\hat{\theta}_2$ for a parameter Θ, we consider $\hat{\theta}_1$ to be better than $\hat{\theta}_2$ if $\mathrm{MSE}(\hat{\theta}_1) \leq \mathrm{MSE}(\hat{\theta}_3)$.

Let us consider an example. Suppose $X_1, X_2, \cdots, X_n$ are $n > 1$ independent random observations taken from a distribution with mean μ and finite variance σ^2. Let us consider some estimators for μ,

$$\hat{\mu}_1 = \bar{X} = \frac{X_1 + X_2 + \cdots + X_n}{n}, \quad \hat{\mu}_2 = \frac{2X_1 + X_2}{3}$$

$$\text{and} \quad \hat{\mu}_3 = \frac{2X_1 + X_2}{4}.$$

It is easy to show that $E(\hat{\mu}_1) = \mu$, $E(\hat{\mu}_2) = \mu$ and $E(\hat{\mu}_3) = 0.75\,\mu$. Thus $\hat{\mu}_1$ and $\hat{\mu}_2$ are unbiased estimators of μ but $\hat{\mu}_3$ is a biased estimator. Using the MSE criteria, how do these estimators compare?

$$\mathrm{MSE}(\hat{\mu}_1) = \mathrm{Var}(\bar{X}) + E(E(\bar{X}) - \mu)^2 = \mathrm{Var}(\bar{X}) + 0 = \frac{\sigma^2}{n},$$

$$\mathrm{MSE}(\hat{\mu}_2) = \mathrm{Var}(\hat{\mu}_2) + E(E(\hat{\mu}_2) - \mu)^2 = \mathrm{Var}(\hat{\mu}_2) + 0 = \frac{5\sigma^2}{9},$$

$$\mathrm{MSE}(\hat{\mu}_3) = \mathrm{Var}(\hat{\mu}_3) + E(E(\hat{\mu}_3) - \mu)^2 = \frac{5\sigma^2}{16} + E(0.75\mu - \mu)^2$$

$$= \frac{5\sigma^2}{16} + \frac{\mu^2}{16}.$$

$\mathrm{MSE}(\hat{\mu}_1) < \mathrm{MSE}(\hat{\mu}_2)$ and $\mathrm{MSE}(\hat{\mu}_1) < \mathrm{MSE}(\hat{\mu}_3)$ and thus the sample mean $\bar{X} = \hat{\mu}_1$ is the best estimator among the three estimators using the MSE criterion. Between $\hat{\mu}_2$ and $\hat{\mu}_3$, which is the better estimator? It depends on the magnitude of σ and μ. If $35\sigma^2 < 9\mu^2$, then $\mathrm{MSE}(\hat{\mu}_2) < \mathrm{MSE}(\hat{\mu}_3)$ and $\hat{\mu}_2$ is a better estimator than $\hat{\mu}_3$, otherwise, it is the other way round. A point of this example is that sometimes a biased estimator such as $\hat{\mu}_3$ can be better than an unbiased estimator such as $\hat{\mu}_2$ using the *MSE* criterion especially when the bias is not too large.

4.2.7 *Summary*

To construct CI estimates for population mean μ, we distinguish between two cases. In both cases, we assume that the population variance σ^2 is finite.

Case 1: The population has the normal distribution $N(\mu, \sigma^2)$. When σ is known, a $1 - \alpha$ CI for μ is given by:

$$(\bar{X} - Z_{1-\alpha/2} \cdot \sigma/\sqrt{n}, \bar{X} + Z_{1-\alpha/2} \cdot \sigma/\sqrt{n}).$$

When σ is unknown, we estimate σ using the sample standard deviation s. In this case, a $1 - \alpha$ CI for μ is given by

$$(\bar{X} - t_{n-1}(1 - \alpha/2) \cdot s/\sqrt{n}, \bar{X} + t_{n-1}(1 - \alpha/2) \cdot s/\sqrt{n}).$$

Note that when n is large, the t_{n-1} distribution is very close to the normal distribution.

Case 2: The distribution for the population is unknown. The sample size is large, usually 30 observations or more.

When σ is known, a $1 - \alpha$ CI for μ is given by

$$(\bar{X} - Z_{1-\alpha/2} \cdot \sigma/\sqrt{n}, \bar{X} + Z_{1-\alpha/2} \cdot \sigma/\sqrt{n}).$$

When σ is unknown, we estimate σ using the sample standard deviation s. In this case, a $1 - \alpha$ CI for μ is given by:

$$(\bar{X} - Z_{1-\alpha/2} \cdot s/\sqrt{n}, \bar{X} + Z_{1-\alpha/2} \cdot s/\sqrt{n})$$

In most practical situations, we don't know σ.

Choosing a sample size for the estimation of population mean μ using the sample mean $\bar{X}$.

The sample size is always a concern when designing a statistical study. Too large a sample is wasteful in terms of time and money. But, too small a sample makes the conclusion less "reliable". The choice of the sample size required is based on the following:

1. The desired level of confidence.
2. The maximum allowable error, E, the researcher will tolerate.
3. The variability of the population under study (measured by σ).

For a given level of confidence, say $1 - \alpha$ where α is usually small, we want

$$P(|\text{Error}| < E) = P(|\bar{X} - \mu| \leq E) = 1 - \alpha.$$

Note that, $P(|\bar{X} - \mu| \leq E) = P\left(\frac{|\bar{X} - \mu|}{\sigma/\sqrt{n}} \leq \frac{E}{\sigma/\sqrt{n}}\right) = P\left(|Z| \leq \frac{\sqrt{n}E}{\sigma}\right) = 1 - \alpha$ where $Z \sim N(0, 1)$. We see that

$$\frac{\sqrt{n}E}{\sigma} = Z_{1-\alpha/2}$$

and the sample size required is $n = \left(\frac{\sigma Z_{1-\alpha/2}}{E}\right)^2$ where $P(|Z| \leq Z_{1-\alpha/2}) = 1 - \alpha$.

A population with considerable variability σ will require a larger sample. E is the maximum allowable error that you are willing to accept. It is the amount that is added to and subtracted from the sample mean $\bar{X}$ to obtain the endpoints of the $1 - \alpha$ CI.

Example: Coke vs. Pepsi survey revisited

Recall the marketing company that wants to estimate the proportion p of adults in Singapore who prefer Coca-Cola to Pepsi Cola. Suppose the marketing manager sets a requirement that we have to be 95% confident that the error in the estimate will be at most 1%. What is the smallest sample size needed?

Here maximum error size $E = 1\%$ and confidence level $1 - \alpha = 95\%$. We do not know σ for this problem. First, we note that the larger the value of σ, the larger the required sample size will be. So we can do something conservative here by estimating an upper bound for σ. We noted earlier that $\sigma^2 = p(1 - p) \leq \left(\frac{1}{2}\right)^2$ and so $\sigma \leq 0.5$. For the 95% confidence level, we noted that $Z_{0.975} = 1.96$. With all these ingredients, we see that the minimum sample size needed is:

$$n = \left(\frac{\sigma Z_{1-\alpha/2}}{E}\right)^2 = \left(\frac{0.5 \times 1.96}{1\%}\right)^2 = 9{,}604.$$

With this sample size, the confidence level is probably higher than 95% as we have used an upper bound for the unknown σ.

Often, the population standard deviation, σ, has to be estimated before doing the full survey or study to provide some clue about the sample size needed. The following methods of estimation are suggested:

1. Use a comparable study, where an estimate of dispersion in the population is available.
2. Use a range-based approach. If we know, or can estimate the largest and smallest values in the population we can compute the range, R. The standard deviation can then be estimated as $R/6$.
3. Conduct a pilot study. Take a small sample and estimate σ, by calculating s.

Bibliography

Illowsky, B. and Dean, S. (2013). *Introductory Statistics*. OpenStax. Retrieved from openstar.org.

Triola, M. (2018). *Elementary Statistics*. 13th ed., Chaps. 6 and 7. Pearson, New Jersey, NJ.

4.3 Sample Questions

Please select the most appropriate response.

Question 1

A sample of size n is drawn from a population that is known to be normally distributed as $N(\mu, 2^2)$. If we want to estimate the population mean μ to within an accuracy of 1 decimal place with a probability of at least 95%, how large a sample size should we take? Please select the smallest number from the choices below that is adequate for the task.

(a) 30
(b) 900
(c) 1,600

The following information applies to Questions 2 to 6.

Aqil is an analyst with a large fund management company. He created a group of 20 investment instruments using various asset classes and securities from different sectors. All the 20 instruments have similar risk–reward profiles. The annual return on each of them may be modeled as random variables having the normal distribution $N(10\%, (30\%)^2)$. That is, each investment instrument has an average (continuous) return of 10% per year with a standard deviation (volatility) of 30%. Moreover, these instruments are constructed so that their returns may be regarded as independent random variables. He formed three baskets using these investment instruments where each investment instrument is equally weighted in the basket.

Basket A: contains five investment instruments, where 20% of the money is invested in each instrument.

Basket B: contains ten investment instruments, where 10% of the money is invested in each instrument.

Basket C: contains all twenty investment instruments, where 5% of the money is invested in each instrument.

As the weights within each basket are equal, it is easy to see that the annual returns of baskets A, B, and C will be random variables that are normally distributed as

$$N\left(10\%, \frac{(30\%)^2}{5}\right), \quad N\left(10\%, \frac{(30\%)^2}{10}\right) \quad \text{and} \quad N\left(10\%, \frac{(30\%)^2}{20}\right)$$

respectively.

Alice is a sophisticated investor who is interested in these baskets.

Question 2

Which basket will give her the highest chance of getting a return of 10% or higher in a year?

(a) Basket B
(b) Basket C
(c) None of the above. They all have the same chance of delivering a return of 10% or higher in a year.

Question 3

Which basket will give her the highest chance of getting a return of 9% or higher in a year?

(a) Basket A
(b) Basket B
(c) Basket C

Question 4

Which basket will give Alice the highest chance of getting a return of 12% or higher in a year?

(a) Basket A
(b) Basket B
(c) Basket C

Question 5

Having heard from her friends that diversification is a good principle for investments, Alice decided to invest in basket C. What is the chance that basket C will lose money, that is, have a negative return, at the end of one year?

(a) 4.50%
(b) 6.80%
(c) 8.30%

Question 6

As Alice is a long-term investor, she is prepared to hold onto basket C for many years. The return on a basket in a year is independent of the returns in subsequent years. She learned from YouTube talks that the longer she holds onto her portfolio the higher the chance that the average return per year over the period will be close to the mean of 10%. How many years should she hold basket C so that there is an 80% chance that the average return per year over the holding period is at least 9%? Please provide an answer to the nearest number of years.

(a) 15 years
(b) 32 years
(c) 38 years

Question 7

A swimming program was designed to help children in the 10- to 12-year age group improve the time taken to complete a 100 m swim. A study of 50 swimmers who completed the program found that the average improvement in the sample was 2.9 seconds with the sample standard deviation being 4.2 seconds. Which is a 95% CI for the average improvement in the population?

(a) (1.74, 4.06)
(b) (1.78, 4.02)
(c) (1.92, 3.88)

Question 8

A swimming program was designed to help children in the 10- to 12-year age group improve the time taken to complete a 100 m swim. For a child, the improvement in the time taken to complete a 100 m swim may be considered a random variable having a normal distribution. A study of 15 swimmers who completed the program found that the average improvement in the sample was 2.9 seconds with the sample standard deviation being 4.2 seconds. Which is a 95% CI for the average improvement in the population?

(a) (0.99, 4.81)
(b) (0.59, 5.21)
(c) (0.57, 5.23)

The following information applies to Questions 3 to 5.

To obtain an estimate of the average claim for a certain medical treatment, an insurance company reviewed a total of 70 medical records in two batches. The first batch of 30 records has a sample mean of $2,813 with a sample standard deviation $230. The second batch of 40 records has a sample mean of $2,758 with a sample deviation of $393.

Question 9

Please provide an estimate for the average claim in the population.

(a) $2,758.00
(b) $2,781.57
(c) $2,785.50

Question 10

Please provide an estimate for the standard deviation of the population of claims.

(a) $230.00
(b) $311.50
(c) $332.09

Question 11

If you need a 95% CI for the average claim in the population, which interval would you use?

(a) ($2,636.21, $2,879.79)
(b) ($2,682.30, $2,888.70)
(c) ($2,703.78, $2,859.37)

Solutions

Question 1

Solution: Option **c** is correct.

The sample mean $\bar{X}$ has the normal distribution $N\left(\mu, \frac{2^2}{n}\right)$. To have an accuracy of one decimal place with a 95% chance, we typically pick two standard deviations to be equal to or less than 0.1. We set $2 \times \frac{2}{\sqrt{n}} \leq 0.1$. Solving gives $1{,}600 \leq n$.

Question 2

Solution: Option **c** is preferred though they are all correct theoretically.

Since their returns are normally distributed with the same mean of 10%, each has a 50% chance of returning 10% or higher.

Question 3

Solution: Option **c** is correct

The average returns are normally distributed with the same mean of 10%. The distribution having the smallest left tail is the one with the smallest standard deviation (more concentrated around the mean of 10%). That is, basket C has the smallest probability of a return less than 9%.

Question 4

Solution: Option **a** is correct.

The average returns are normally distributed with the same mean of 10%. The distribution having the largest right tail is the one with the largest standard deviation. That is, basket A has the highest probability of a return of 12% or higher. Another way to think about this is to note that the bigger the basket the harder it is for the return on the basket (sample mean) to be far away from the mean of 10%. A smaller basket has returns that are more variable and hence a higher chance of getting 12% or higher.

Question 5

Solution: Option **b** is correct.

The one-year return, R, on basket C has the normal distribution $N\left(10\%, \frac{(30\%)^2}{20}\right)$. Thus $P(R < 0\%) = P\left(\frac{R-10\%}{30\%/\sqrt{20}} < \frac{0\%-10\%}{30\%/\sqrt{20}}\right) = P(Z < -1.4907) = 6.80\%$.

Question 6

Solution: Option **b** is correct.

Consider holding basket C for n years. The average return $\bar{R}$ per year over the holding period has the normal distribution $N\left(10\%, \frac{30\%^2/20}{n}\right)$. We want to determine n so that $P(\bar{R} \geq 9\%) = 0.80$. Thus

$$P(\bar{R} \geq 9\%) = P\left(\frac{\sqrt{n}(\bar{R} - 10\%)}{\sqrt{(30\%)^2/20}} \geq \frac{\sqrt{n}(9\% - 10\%)}{\sqrt{(30\%)^2/20}}\right)$$

$$= P(Z \geq -1.4907\sqrt{n}) = 0.8.$$

From the tables, $-1.4907\sqrt{n} = -0.84162$. Solving gives $n = 31.9$ years. This example shows that volatility is detrimental to investment returns. Even when Alice wants an average return of just 9% or higher, she has to hold the basket for 32 years to have a probability of 80% of attaining the objective.

Question 7

Solution: Option **a** is correct.

Using the central limit theorem, the sample mean $\bar{X}$ follows a normal distribution $N(\mu, \frac{s^2}{n})$ approximately, where sample size $n = 50$ and sample standard deviation $s = 4.2$. A 95% CI for μ has endpoints given by

$$\bar{X} \pm Z_{0.025}\frac{s}{\sqrt{n}} = 2.9 \mp 1.96 \times \frac{4.2}{\sqrt{50}} = 1.74, 4.06.$$

Question 8

Solution: Option **c** is correct.

Let μ and $\bar{X}$ be the population mean and sample mean, respectively. Let s be the sample standard deviation and n be the sample size. We recall that $T = \frac{\bar{X}-\mu}{s/\sqrt{n}} \sim t_{n-1}$. A 95% CI for μ has endpoints given by

$$\bar{X} \pm t_{n-1}(0.025)\frac{s}{\sqrt{n}} = 2.9 \mp 2.1448 \times \frac{4.2}{\sqrt{15}} = 0.57, 5.23.$$

Question 9

Solution: Option **b** is correct.

The sample means are good estimates of the population mean. Can we combine them to arrive at better estimates? Yes, a simple arithmetic average $= (2{,}813 + 2{,}758)/2 = \$2{,}785.50$ may be a reasonable estimate. Can we obtain a "better" estimate than the simple average of the two? Yes, a better estimate is to weigh each observation equally. That is, a better estimate is a weighted average of the two sample means, namely $(30 \times 2{,}813 + 40 \times 2{,}758)/70 = \$2{,}781.57$.

Question 10

Solution: Option **c** is correct.

There are various estimates for the standard deviation of the population. A reasonable estimate is to combine the two batches into a big sample and compute the standard deviation of the big sample. Let n_i, $\bar{X}_i$, and s_i be the sample size, the sample mean, and sample standard deviation respectively for batch i with $i = 1$ for the first batch of 30 records and $i = 2$ for the second batch of 40 records. Let $\bar{X}$ and s be the sample mean and sample standard deviation of the combined sample of 70 records.

We note that $\bar{X} = \frac{n_1\bar{X}_1+n_2\bar{X}_2}{n_1+n_2} = \frac{30\times2{,}813+40\times2{,}758}{30+40} = 2{,}781.57$. The sample standard deviation for the combined sample is given by

$$s = \sqrt{\frac{1}{n_1 + n_2 - 1}[(n_1 - 1)s_1^2 + (n_2 - 1)s_2^2 + \frac{n_1 n_2}{n_1 + n_2}(\bar{X}_1 - \bar{X}_2)^2]}$$
$$= 332.09.$$

Question 11

Solution: Option **c** is correct.

All the confidence intervals are reasonable. The most efficient one is (d) as it has a 95% confidence level and the shortest interval among the four. It is constructed using the sample mean $\bar{X}$ and sample standard deviation s for the combined sample of 70 observations, namely $\bar{X} = 2{,}781.57$ and $s = 332.09$. The endpoints for a 95% confidence level for the population mean are $\bar{X} \pm z_{0.975}\frac{s}{\sqrt{n_1+n_2}} = 2{,}781.57 \pm 1.96\frac{332.09}{\sqrt{70}}$.

Chapter 5

Hypothesis Testing

5.1 Introduction and Hypothesis Testing

Hypothesis testing is an act in statistics whereby an analyst tests an assumption regarding a population parameter.

5.1.1 *Learning Objectives*

- Define a hypothesis, steps of hypothesis testing, and interpret the choice of the null and alternative hypotheses.

5.1.2 *Main Takeaways*

5.1.2.1 *Main Points*

- In reaching a statistical decision, we can make two possible errors: We may reject a true null hypothesis (a Type I error), or we may fail to reject a false null hypothesis (a Type II error).
- If we are applying a one-tailed test, there is one critical value. If we are applying a two-tailed test, there are two critical values.
- We state two hypotheses: The null hypothesis is the hypothesis to be tested; the alternative hypothesis is the hypothesis accepted when the null is rejected.

5.1.2.2   *Main Terms*

- **Test statistics:** A quantity is calculated based on a sample, whose value is the basis for deciding whether or not to reject the null hypothesis.
- **Critical value:** A rejection point (critical value) for a test statistic is a value with which the computed test statistics are compared to decide whether to reject or not reject the null hypothesis.
- **Level of significance:** Also denoted as alpha or α, this is the probability of rejecting the null hypothesis when it is true. For example, a significance level of 0.05 indicates a 5% risk of concluding that a difference exists when there is no actual difference.

5.1.3   *Introduction*

The hypothesis involves testing claims about the characteristics of one or more populations using samples drawn from them. The strength of the evidence found in the sample will be used to determine whether or not the hypothesis will be rejected. The chance that an erroneous conclusion is obtained due to the vagaries of sampling is controlled by setting the level of significance of the test.

We discuss the steps involved in the testing of a hypothesis, paying particular attention to the concept of the level of significance and its connection to the types of errors in a test.

A hypothesis is a statement about the properties of one or more populations. For example, a researcher may wish to determine whether a medical treatment or a public policy is effective. Hypothesis testing allows us to come to a conclusion based on the strength of the evidence found in a dataset or sample. For example, a car company claims that the Model Y electric cars that it developed can travel 350 km on a full charge. The process of verifying a claim may be carried out using the following steps shown in Table 5.1.

Let us use the car company's claim as an example. Let μ denote the average mileage in km of the Model Y electric cars on a full charge.

Null hypothesis, H_0: $\mu = 350$,
Alternative hypothesis, H_{A1}: $\mu \neq 350$.

This setup is known as a two-tailed test. We are going to use the average mileage $\bar{X}$ in the sample (sample mean) to test this claim.

Table 5.1: Five-step process to hypothesis testing.

Step 1	Step 2	Step 3	Step 4	Step 5
State the null and alternative hypotheses	Select a level of significance	Identify a test statistic	Formulate a decision rule based on the sampling distribution of the test statistic under the null hypothesis	Take a sample. Decide on the rejection or otherwise of the null hypothesis based on the value of the test statistic

When the sample mean $\bar{X}$ is close to 350 km, we do not reject the null hypothesis. However, if the sample mean is far below or far above the stipulated value of 350 km, the null hypothesis will be rejected. There are two regions $\bar{X} \ll 350$ ($\bar{X}$ is much lower than 350) and $\bar{X} \gg 350$ ($\bar{X}$ is much higher than 350) where the null hypothesis will be rejected. Hence, this is known as a two-tailed test. In this example, naturally, a consumer is only concerned when the average mileage μ is lower than 350 km. Consumers are happy if the average mileage is above 350 km.

If only one side of 350 km causes concern, we may specify the null and alternative hypothesis as follows:

Null hypothesis, H_0: $\mu = 350$,
Alternative hypothesis, H_{A2}: $\mu < 350$.

This setup is a one-tailed test where we reject the null hypothesis and accept the alternative when the sample mean $\bar{X}$ is much lower than 350 km. There is only one rejection region for the null hypothesis, namely when $\bar{X} \ll 350$, that is, when $\bar{X}$ is much lower than 350 km.

How about setting up the hypotheses as follows?

Null hypothesis, H_0: $\mu \geq 350$,
Alternative hypothesis, H_A: $\mu < 350$.

This is a reasonable setup where the null hypothesis says that the average mileage is 350 km or more. This type of null hypothesis is known as a composite null hypothesis where the null hypothesis comprises more than one probability distribution. For example, the null hypothesis $\mu \geq 350$ includes the cases $\mu = 350$ or $\mu = 375$ or $\mu = 409.2$. A composite null hypothesis can make the test more complicated but not in this instance. We are lucky, in this example, in that it suffices to consider the simple null hypothesis, H_0: $\mu = 350$. If the null hypothesis is rejected for $\mu = 350$, then it will be rejected for all null hypotheses $\mu = M$ with $M \geq 350$.

5.1.4 *Two-tailed Test or Two-sided Test*

Let us consider the two-tailed test of the hypothesis.

Step 1: State the null and alternative hypotheses

Null hypothesis, H_0: $\mu = 350$,
Alternative hypothesis, H_{A1}: $\mu \neq 350$.

Step 2: Select a level of significance

The level of significance $\alpha = 5\%$. In other words, if an event has a chance of 5% or less of occurring, we consider this to be a significant or rare event. There is nothing magical about 5%, some may use 1% or 2% as the level of significance. This has to be set before we look at the data to maintain objectivity. We should not look at the data and then decide on the level of significance to achieve the desired outcome.

Step 3: Identify the test statistic

We are going to charge up n Model Y electric cars fully and then measure their mileage $X_1, X_2, \ldots, X_n$. The sample mean $\bar{X} = (X_1 + X_2 + \cdots + X_n)/n$ is an unbiased estimator for the population mean μ. So, naturally, the sample mean $\bar{X}$ can be used as a test statistic.

Step 4: Formulate a decision rule

In the hypothesis testing framework, we assume that the null hypothesis is valid as a default when we do the test. We need to determine the sampling distribution of the test statistic $\bar{X}$, assuming that the null hypothesis is true, that is, assuming that $\mu = 350$. If the sample size n is large ($n \geq 30$), we can invoke the Central Limit Theorem to say that the sampling distribution of $\bar{X}$ is approximately the normal distribution $N\left(\mu, \frac{\sigma^2}{n}\right)$ where σ^2 is the population variance. If the population variance is not known, we estimate it using the sample variance $s^2 = \frac{\sum_{i=1}^{n}(X_i - \bar{X})^2}{n-1}$.

Under the null hypothesis, $\bar{X} \sim N\left(\mu, \frac{s^2}{n}\right)$ approximately. If the null hypothesis is valid, then we are sampling a $\bar{X}$ from this distribution and we should get an observation near μ. The rejection region is determined by the level of significance α. This is a two-tailed test with two rejection regions being the values that are way below and way above μ. We divide the level of significance equally into two and the rejection regions each is associated with a probability of $\alpha/2 = 2.5\%$.

Thus, the rejection regions are $\bar{X} \leq \mu + Z_{\alpha/2}\frac{s}{\sqrt{n}}$ and $\bar{X} \geq \mu + Z_{1-\alpha/2}\frac{s}{\sqrt{n}}$ where $Z \sim N(0,1)$ and $P(Z \leq Z_{\alpha/2}) = \frac{\alpha}{2}$ and $P(Z \geq Z_{1-\alpha/2}) = \frac{\alpha}{2}$. With $\alpha = 5\%$, the decision rule is to reject the null hypothesis when $\bar{X} \leq 350 - 1.96\frac{s}{\sqrt{n}}$ or $\bar{X} \geq 350 + \frac{s}{\sqrt{n}}$.

We may convert the sample mean into a standardized Z score as follows:

$Z = \frac{\bar{X}-\mu}{s/\sqrt{n}} \sim N(0,1)$ and use it as a test statistic where the rejection regions are given by $Z \leq Z_{\alpha/2} = -1.96$ and $Z \geq Z_{1-\alpha/2} = 1.96$.

The rejection region is also known as the critical region and the values ± 1.96, in this case, are known as the critical values.

Step 5: Obtain experimental data and make a conclusion

Suppose we run an experiment with 45 Model Y cars on a full charge and find that the average mileage achieved was 344.9 km with a sample standard deviation of 18.3 km.

Here $n = 45$, $\bar{X} = 344.9$ km and $s = 18.3$ km. The rejection regions are $\bar{X} \leq 344.65$ and $\bar{X} \geq 355.35$. Since $\bar{X} = 344.9$, we do not reject the null hypothesis.

We may also use $Z = \frac{\bar{X}-\mu}{s/\sqrt{n}} = \frac{344.9 - 350}{18.3/\sqrt{45}} = -1.8695$ to make a decision. The rejection regions for Z are $Z \leq -1.96$ and $Z \geq 1.96$. Since $Z = -1.8695$ is outside the rejection regions, the null hypothesis is not rejected.

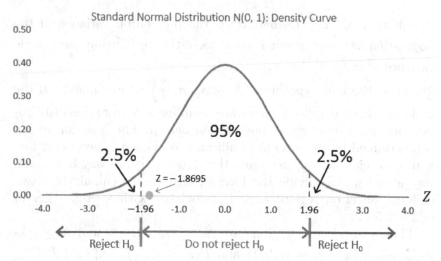

Figure 5.1: Standard normal distribution density curve for 95% confidence interval.

5.1.5   *Not Rejecting the Null Hypothesis is Not the Same as Believing that it is True*

When the null hypothesis is not rejected, we sometimes say that "the null hypothesis is accepted". Though these two clauses may be equivalent logically, it is useful to keep in mind that when we say "the null hypothesis is accepted", it is not that we believe that the null hypothesis is true, but rather there is insufficient evidence at the moment to reject it.

Figure 5.1 shows pictorially the density curve for the Z score and the rejection regions at the two tails of the distribution.

5.1.6   *The p-value of the Test*

In the test, the level of significance seems to be arbitrarily chosen. If we choose 7% as the level of significance in the test involving the Model Y cars, we would have rejected the null hypothesis as the rejection regions are $\bar{X} \leq 345.06$ and $\bar{X} \geq 354.94$. To overcome the arbitrariness of picking the level of significance, it is customary to report the p-value of the test. The p-value of the test is the probability, under the null hypothesis, of obtaining the test statistic's observed value or something more severe against the null hypothesis.

Under the null hypothesis $\bar{X} \sim N(\mu, \frac{s^2}{n}) = N(350, \frac{18.3^2}{45})$, and the probability of obtaining the test statistic of $\bar{X} = 344.9$ or something more severe against the null hypothesis is

$$P(\bar{X} \leq 344.9) = P\left(\frac{\bar{X} - \mu}{s/\sqrt{n}} \leq \frac{344.9 - 350}{18.3/\sqrt{45}}\right)$$
$$= P(Z \leq -1.8695) = 0.0308.$$

As this is a two-tailed test, the p-value of the test is $2 \times 0.0308 = 6.16\%$. That is, if we set the level of significance to be higher than 6.16%, we will reject the null hypothesis. We see that if the p-value of the test is lower than the level of significance, the null hypothesis will be rejected. The smaller the p-value, the stronger is the evidence against the null hypothesis. Reporting the p-value gives us a sense of the strength of the evidence against the null hypothesis.

5.1.7 The Connection Between Hypothesis Testing and Confidence Intervals

In the test of the hypothesis concerning the population mean of the mileage of Model Y electric cars, we used a sample of 45 cars and obtained a sample mean of 344.9 km with a sample standard of 18.3 km. The level of significance was $\alpha = 5\%$. We can use the data from the sample to construct a $1 - \alpha = 95\%$ confidence interval for the population mean μ. The symmetric 95% confidence interval has endpoints given by

$$\bar{X} \pm Z_{\alpha/2} \times \frac{s}{\sqrt{n}} = 344.9 \mp 1.96 \times \frac{18.3}{\sqrt{45}} = 339.55 \text{ and } 350.25.$$

Thus, a 95% confidence interval for μ is (339.55, 350.25). Note that the value $\mu = 350$ is inside the interval. This means that the two-tailed test of the null hypothesis $\mu = 350$ will not be rejected at the 5% level of significance. On the other hand, the two-tailed test of the null hypothesis $\mu = 351$ will be rejected at the 5% level of significance as the value $\mu = 351$ is not inside the confidence interval.

5.1.8   *One-tailed Test or One-sided Test*

Since consumers are concerned about low mileage rather than high mileage, it makes perfect sense to do a one-tailed test as follows:

Null hypothesis, H_0: $\mu = 350$,
Alternative hypothesis, H_{A2}: $\mu < 350$,
Level of significance, $\alpha = 5\%$.

We use the same dataset as the two-tailed test with $n = 45$, $\bar{X} = 344.9$ and $s = 18.3$.

The rejection region will be $\bar{X} \le \mu + Z_\alpha \frac{s}{\sqrt{n}} = 350 - 1.6449 \times \frac{18.3}{\sqrt{45}} = 345.51$.

Since the observed test statistic $\bar{X} = 344.9$ which falls inside the rejection region, we reject the null hypothesis and accept the alternative hypothesis that $\mu < 350$.

We can also use the Z score as a test statistic where $Z = \frac{\bar{X} - \mu}{s/\sqrt{n}} \sim N(0,1)$.

The rejection region is $Z < Z_\alpha = -1.6449$.

From the sample data, we compute $Z = \frac{\bar{X} - \mu}{s/\sqrt{n}} = \frac{344.9 - 350}{18.3/\sqrt{45}} = -1.8695$. We arrive at the same conclusion, namely to reject the null hypothesis.

What is the p-value of the test? That is, under the null hypothesis, what is $P(\bar{X} \le 344.9)$?

Recall that $P(\bar{X} \le 344.9) = P\left(\frac{\bar{X}-\mu}{s/\sqrt{n}} \le \frac{344.9 - 350}{18.3/\sqrt{45}}\right) = P(Z \le -1.8695) = 0.0308$. Thus, the p-value of the test is 3.08%. Since the p-value is less than the level of significance of 5%, we see that the null hypothesis is rejected.

The rejection region, critical value, and the p-value are illustrated in Figure 5.2.

Let us again draw the connection between hypothesis testing and confidence interval for the population mean μ. With the data, let us construct a one-sided confidence interval for μ. A one-sided $1 - \alpha = 95\%$ confidence interval for μ is $(-\infty, \bar{X} + Z_{1-\alpha} \times \frac{s}{\sqrt{n}}) = (-\infty, 349.39)$. Since this one-sided confidence interval does not contain $\mu = 350$, the one-tailed test of the null hypothesis will be rejected at the 5% level of significance.

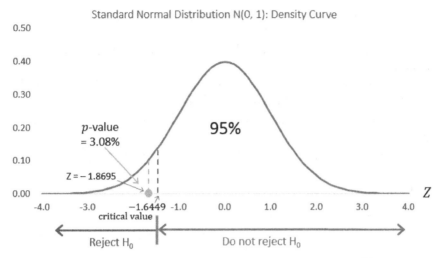

Figure 5.2: Standard normal distribution density curve for 95% confidence interval.

5.1.9 *Can We Handle a Small Sample Size?*

The process for testing a hypothesis relies on knowing the sampling distribution of the test statistic. Suppose it is very expensive or inconvenient to collect the data for Model Y electric cars and we have a small sample size of 9. The test statistic is again the sample mean. But we do not know the sampling distribution of the sample mean $\bar{X}$ since we cannot invoke the Central Limit Theorem. We are not able to proceed unless we make some assumptions about the population. Suppose we assume that the mileage of a Model Y electric car on a full charge follows a normal distribution $N(\mu, \sigma^2)$. This assumption is sufficient for us to derive the sampling distribution of the test statistic $\bar{X}$. Let n be the sample size and s be the sample standard deviation. It follows that $T = \frac{\bar{X}-\mu}{s/\sqrt{n}} \sim t_{n-1}$ has the t-distribution with $n-1$ degrees of freedom. We can then proceed to test the hypothesis using this fact.

5.1.10 *One-tailed Test with a Small Sample Size*

Null hypothesis, H_0: $\mu = 350$,
Alternative hypothesis, H_{A2}: $\mu < 350$,
Level of significance, $\alpha = 5\%$.

Assumption: The mileage of a Model Y electric car on a full charge is normally distributed as $N(\mu, \sigma^2)$.

Suppose we collected the following information from a sample of 9 cars.

Sample size $n = 9$, the sample mean $\bar{X} = 346.1$ km, sample standard deviation $s = 35$ km.

Then $T = \frac{\bar{X}-\mu}{s/\sqrt{n}} \sim t_{n-1} = t_8$ has the t-distribution with 8 degrees of freedom.

The rejection region for this one-sided test is $T \leq t_8(0.05) = -1.8596$.

From the sample,

$$T = \frac{\bar{X}-\mu}{s/\sqrt{n}} = \frac{346.1 - 350}{35/\sqrt{9}} = -0.3343 > t_8(0.05) = -1.8596.$$

Conclusion: The null hypothesis is not rejected. The p-value of the test is 37.34%.

Besides testing claims about population parameters such as mean and variance, or proportion or quantile of a distribution, the framework for hypothesis testing may be used to check whether a population has a specified distribution in a goodness-of-fit test. It may be applied to testing claims about two or more populations. The key is to find suitable test statistics whose sampling distributions are known exactly or approximately under the null hypothesis.

5.1.11 *Hypothesis Testing: Difference Between Two Population Means*

Let us keep the Model Y electric car example. Suppose a rival company has a Model A electric car that has the same energy capacity as the Model Y car. The rival company claims that the Model A car has higher mileage than a Model Y car on average. The hypothesis testing framework may be applied to test such claims. Let μ_A and μ_Y denote the average mileages of the Model A and Model Y electric cars, respectively, on a full charge.

We wish to test the following hypothesis:

Null hypothesis, H_0: $\mu_A = \mu_Y$,
Alternative hypothesis, H_A: $\mu_A > \mu_Y$,
Level of significance $\alpha = 5\%$.

We will collect the mileages on a group of Model A and Model Y cars and obtain the sample averages of the mileages for the two models.

Let n_A and n_Y be the number of observations for the Model A and Model Y cars, respectively.

Let $\bar{X}_A$ and $\bar{X}_Y$ be the sample averages of the mileages for Model A and Model Y cars, respectively.

Let s_A and s_Y be the sample standard deviations for Model A and Model Y cars, respectively.

If both sample sizes n_A and n_Y are large enough, we have

$$\bar{X}_A \sim N\left(\mu_A, \frac{\sigma_A^2}{n_A}\right) \text{ and } \bar{X}_Y \sim N\left(\mu_Y, \frac{\sigma_Y^2}{n_Y}\right) \text{ approximately.}$$

Given that the observations in the samples are independent of each other, $\bar{X}_A - \bar{X}_Y$ will be normally distributed as

$$\bar{X}_A - \bar{X}_Y \sim N\left(\mu_A - \mu_Y, \frac{\sigma_A^2}{n_A} + \frac{\sigma_Y^2}{n_Y}\right).$$

Recall that $\text{Var}(\bar{X}_A - \bar{X}_Y) = \text{Var}(\bar{X}_A) + (-1)^2 \text{Var}(\bar{X}_Y) = \frac{\sigma_A^2}{n_A} + \frac{\sigma_Y^2}{n_Y}$ when the two sample means are independent.

We may use $\bar{X}_A - \bar{X}_Y$ as the test statistic or we could use the Z score as the test statistic, where

$$Z = \frac{(\bar{X}_A - \bar{X}_Y) - (\mu_A - \mu_Y)}{\sqrt{\frac{\sigma_A^2}{n_A} + \frac{\sigma_Y^2}{n_Y}}} \sim N(0, 1).$$

Under the null hypothesis, $\mu_A = \mu_Y$ and so,

$$Z = \frac{(\bar{X}_A - \bar{X}_Y) - (\mu_A - \mu_Y)}{\sqrt{\frac{\sigma_A^2}{n_A} + \frac{\sigma_Y^2}{n_Y}}} = \frac{\bar{X}_A - \bar{X}_Y}{\sqrt{\frac{\sigma_A^2}{n_A} + \frac{\sigma_Y^2}{n_Y}}} \sim N(0, 1).$$

The rejection region is given by $Z > Z_{1-\alpha} = Z_{0.95} = 1.6449$.

Suppose the experiment was conducted and Table 5.2 depicts the results.

Using the data, we compute $Z = 1.4031 < Z_{0.95} = 1.6449$ which is not in the rejection region. The p-value of the test is 8.03%.

Conclusion: The evidence is not strong enough to reject the null hypothesis. The data does not support the rival company's claim that the average mileage of the Model A car is higher than that of the Model Y car.

Table 5.2: Sample size, sample mileage and sample standard deviations of Model A and Y.

	Model A	Model Y
Sample size	32	45
Sample mean mileage	352.7	344.9
Sample standard deviation	27.4	18.3

Table 5.3: Decision matrix of Type I and Type II error.

		Reality or nature	
		H_0 is true	H_0 is false
Decision	Do not reject H_0	Correct decision	**Type II error**
	Reject H_0	**Type I error**	Correct decision

Notes: Type I error: rejecting the null hypothesis when it is true. Type II error: not rejecting the null hypothesis when it is false.

5.1.12 *Making Mistakes in Hypothesis Testing: Types of Errors*

The procedure for hypothesis testing is such that the null hypothesis H_0 will be rejected once the test statistic falls inside the rejection region. The rejection region is constructed under the assumption that the null hypothesis holds and the chance of the test statistic being in the rejection region is α. We could make two types of errors in hypothesis testing, namely rejecting the null hypothesis H_0 when it is valid or true and failing to reject the null hypothesis H_0 when it is not valid. These errors are named Type I and Type II errors, respectively, as shown in Table 5.3.

Note that the level of significance α is the probability of making a Type I error. In the framework for testing hypothesis, Type I error is usually considered to be the more serious of the two types of errors and hence we set the level of significance to control the chance of making a Type I error. There is a trade-off between Type I and Type II errors. If we reduce the chance of making a Type I error, we are more likely to retain the null hypothesis even when it is false, which means we will increase the chance of making a Type II error. Conversely, if we reduce the chance of a Type II error, we will increase the chance of a Type I error. The way to improve the effectiveness of a test is to collect more observations, that is to have a larger sample.

5.1.13 *Testing Hypothesis: Potential Pitfalls*

Fishing for an atypical sample: Suppose we set the level of significance of the test to be $\alpha = 5\%$. This means that there is a 5% chance that the null hypothesis will be rejected even when it is true. A mischievous researcher goes about testing a true null hypothesis. On many occasions, the correct conclusion of not rejecting the null hypothesis was obtained and these data and attempts were discarded and not disclosed.

One day, by random chance, an atypical sample that resulted in the rejection of the null hypothesis was obtained. This rejection is publicized, thereby giving the incorrect impression that the null hypothesis is not true. With a 5% level of significance, about one in 20 samples or tests will result in a Type I error when the null hypothesis is true.

Framing the alternative hypothesis: Suppose the makers of the Model Y electric cars know that the average mileage μ on a full charge is way below 350 km. How can they do a test so that they can keep the null hypothesis that $\mu = 350$ km? Consider the statement of the hypotheses as follows:

Null hypothesis, H_0: $\mu = 350$,
Alternative hypothesis, H_A: $\mu > 350$.

The rejection region is found much above 350. The sample mean for the mileages of a sample of Model Y cars will likely be below 350 km. This means that the null hypothesis will not be rejected simply because of the way the alternative hypothesis is set up.

Bibliography

Illowsky, B. and Dean, S. (2013). *Introductory Statistics.* OpenStax, pp. 505–595. Retrieved from openstar.org.

Triola, M. (2018). *Elementary Statistics.* 13th ed., Chap. 8. Pearson, New Jersey, NJ.

5.2   Sample Questions

Please select the most appropriate response.

The following information applies to Question 1 and Question 2.

A swimming program is designed to help children in the 10–12 age group improve the time taken to complete a 100 m swim. The coach claims that the average improvement μ in the time taken to complete a 100 m swim will be 3 seconds on average after the program. To test the null hypothesis, H_0: $\mu = 3$ versus the alternative hypothesis, H_A: $\mu < 3$, a sample of 50 swimmers who completed the program were tested. For this sample, the average improvement $\bar{X}$ was 2.1 seconds with the sample standard deviation of 4.5 seconds. The level of significance was set at $\alpha = 5\%$.

Question 1

If the sample mean $\bar{X}$ is used as the test statistic, what is the critical value for this test?

(a) 1.72
(b) 1.75
(c) 1.95

Question 2

What is the p-value of the test?

(a) less than 6%
(b) 7.86%
(c) 15.73%

The following information applies from Question 3 to Question 6.

An insurance company wants to estimate the average size of a claim for certain automobile damages. From a sample of 50 claims, it was found that the sample average was $502 with a sample standard deviation of $49. The consultant working on the project requested additional data to obtain better accuracy in the estimates. A second sample of 30 claims had a sample average of $529 with a sample standard deviation of $70. The consultant suspected that the two samples may not be drawn from the same population and in particular, they come from populations with different population means. Let μ_1 and μ_2 denote the population means association with the first and second samples, respectively. Let $\bar{X}_1$ and $\bar{X}_2$ denote the sample means of the first and second samples, respectively. Let s_1 and s_2 denote the sample standard deviations of the first and second samples, respectively. The consultant wished to test the null hypothesis $H_0: \mu_1 = \mu_2$.

Question 3

Given the data collected, which of the following hypothesis will be a suitable alternative hypothesis for the test?

I. $H_A : \mu_1 > \mu_2$,
II. $H_A : \mu_1 \neq \mu_2$,
III. $H_A : \mu_1 < \mu_2$.

(a) Only alternative hypotheses I and II are suitable.
(b) Only alternative hypotheses I and III are suitable.
(c) Only alternative hypotheses II and III are suitable.

Question 4

Suppose the consultant tested the following hypotheses:

Null hypothesis, $H_0: \mu_1 = \mu_2$,
Alternative hypothesis, $H_A: \mu_1 < \mu_2$,
Level of significance, $\alpha = 5\%$.

What is the critical region for the test?

(a) $\bar{X}_1 - \bar{X}_2 \leq -23.91$
(b) $\bar{X}_1 - \bar{X}_2 \leq -25.45$
(c) $\bar{X}_1 - \bar{X}_2 \leq -28.49$

Question 5

Suppose the consultant tested the following hypotheses:

Null hypothesis, $H_0: \mu_1 = \mu_2$,
Alternative hypothesis, $H_A: \mu_1 \neq \mu_2$,
Level of significance, $\alpha = 5\%$.

What is the acceptance region for the test?

(a) $-23.91 < \bar{X}_1 - \bar{X}_2 < 23.91$
(b) $-25.45 < \bar{X}_1 - \bar{X}_2 < 25.45$
(c) $-28.49 < \bar{X}_1 - \bar{X}_2 < 28.49$

Question 6

Consider the test of the following hypotheses:

Null hypothesis, H_0: $\mu_1 = \mu_2$,
Alternative hypothesis, H_A: $\mu_1 < \mu_2$,
Level of significance, $\alpha = 5\%$.

What is the p-value of the test?

(a) 7.72%
(b) 5.28%
(c) 3.16%

Solutions

Question 1

Solution: Option **c** is correct.

The critical value is $\mu + Z_\alpha \frac{s}{\sqrt{n}} = 3 - 1.6449 \times \frac{4.5}{\sqrt{50}} = 1.95$.

Question 2

Solution: Option **b** is correct

The p-value is $P(\bar{X} \leq 2.1) = P\left(\frac{\bar{X}-\mu}{s/\sqrt{n}} \leq \frac{2.1-3}{4.5/\sqrt{50}}\right) = P(Z \leq -1.41421) = 7.86\%$.

Question 3

Solution: Option **c** is correct

I. This alternative hypothesis does not make sense given the data as the null hypothesis will not be rejected.

I and III are both suitable alternative hypotheses for the test.

Question 4

Solution: Option **a** is correct.

The critical region is

$$\bar{X}_1 - \bar{X}_2 \leq \mu_1 - \mu_2 + Z_\alpha \times \sqrt{\frac{s_1^2}{n_1} + \frac{s_2^2}{n_2}}$$

$$= 0 - 1.6449 \times \sqrt{\frac{49^2}{50} + \frac{70^2}{30}} = -23.91.$$

Question 5

Solution: Option **c** is correct.

The critical values are

$$\mu_1 - \mu_2 \pm Z_{\alpha/2} \times \sqrt{\frac{s_1^2}{n_1} + \frac{s_2^2}{n_2}} = 0 \mp 1.96 \times \sqrt{\frac{49^2}{50} + \frac{70^2}{30}} = \mp 28.49$$

Question 6

Solution: Option **c** is correct.

The *p*-value of the test is

$$P(\bar{X}_1 - \bar{X}_2 \leq 502 - 529)$$

$$= P\left(\frac{(\bar{X}_1 - \bar{X}_2) - (\mu_1 - \mu_2)}{\sqrt{\frac{s_1^2}{n_1} + \frac{s_2^2}{n_2}}} \leq \frac{-27 - 0}{\sqrt{\frac{49^2}{50} + \frac{70^2}{30}}}\right)$$

$$= P(Z \leq -1.8572) = 3.16\%.$$

Chapter 6

Regression

6.1 Multiple Linear Regression

Multiple linear regression is a statistical technique that uses several explanatory variables to predict the outcome of a response variable. Multiple regression is an extension of the linear, or ordinary least-squares (OLS) regression, that uses just one explanatory variable.

6.1.1 *Learning Objectives*

- Formulate a multiple regression equation and interpret the coefficients and regression results.

6.1.2 *Main Takeaways*

6.1.2.1 *Main Points*

- The general form of a multiple linear regression model is $Y_i = b_0 + b_1 X_{1i} + b_2 X_{2i} + \cdots + b_k X_{ki} + \varepsilon_i$.
- The lower the p-value reported for a test, the more significant the result.
- A good grasp of the simple linear model allows us to handle multiple regression with ease as the basic concepts extend to models with more than one independent variable.
- Visual inspection of the pictorial data using scatterplots can help us decide which approach or analysis may be appropriate and it may highlight potential issues that have to be addressed.

- When predicting the dependent variable using a linear regression model, we encounter two types of uncertainty: uncertainty in the regression model itself, as reflected in the standard error of estimate, and uncertainty about the estimates of the regression coefficients.
- R^2 is non-decreasing in the number of independent variables, so it is less reliable as a measure of goodness of fit in regression with more than one independent variable than in a one-independent-variable regression.
- Key information from regression analysis is point and interval estimates for the population parameters. The goal of the model is to predict or forecast the dependent variable Y for a given value of X.
- Linear models may be analyzed by comparing various sums of squared deviations, which is a technique in analysis of variance (ANOVA). In the case of a simple linear model, testing whether the independent variable plays a significant role in the model is the same as performing a test of hypothesis in the ANOVA framework.
- Examining the residuals or error terms from a regression model is an essential step. The residuals may reveal further information or structure that the linear model failed to capture.
- If a regression shows significant conditional heteroskedasticity, the standard errors, and test statistics computed by regression programs will be incorrect unless they are adjusted for heteroskedasticity.
- With multicollinearity, the regression coefficients may not be individually statistically significant even when the overall regression is significant as judged by the F-statistic.
- If a regression is misspecified, then statistical inference using OLS is invalid and the estimated regression coefficients may be inconsistent.

6.1.2.2 *Main Terms*

- **Linear Correlation:** Measures how closely the co-ordinate points formed from two variables are packed around a line.
- **Confidence Interval:** In statistics, a confidence interval (CI) is a type of estimate computed from the statistics of the observed data.

This proposes a range of plausible values for an unknown parameter (for example, the mean).

- **Coefficient of Determination:** A statistical measurement that examines how differences in one variable can be explained by the difference in a second variable, when predicting the outcome of a given event.
- **Confidence Interval:** In statistics, a CI is a type of estimate computed from the statistics of the observed data. This proposes a range of plausible values for an unknown parameter.
- **Residuals:** The difference between the observed value of the dependent variable (y) and the predicted value (y) is called the residual (e). Each data point has one residual.
- **Level of Significance:** The probability of rejecting the null hypothesis when it is true. For example, a significance level of 0.05 indicates a 5% risk of concluding that a difference exists when there *is* no actual difference.

6.1.3 *The Basics*

Linear models are a popular tool in many applications. We study its framework and underlying assumptions, as well as learn how to estimate the parameters of the model and quantify prediction errors.

Linear models are effective tools when they work well. So we need to know when they are applicable and what tools are available to us to draw the best inferences and forecasts.

A linear model allows us to study the relationship among quantities or variables typically labeled $Y, X^1, X^2, \ldots, X^k$. It is believed that the quantity Y is related to, or depends on the variables $X^1, X^2, \ldots, X^k$ in a linear manner, namely $Y = \beta_0 + \beta_1 X^1 + \beta_2 X^2 + \cdots + \beta_k X^k + \varepsilon$ where $\beta_0, \beta_1, \beta_2, \ldots, \beta_k$ are constants that may be estimated from a sample of observations and, ε is a random error. For example, the income Y of a worker may depend on the level of education X^1, the relevant work experience X^2, and other factors. For a given linear model, the following are questions of interest.

- Is the linear model valid? Perhaps, the relationship is best described by non-linear functions.
- How do we measure the efficacy of a linear model?

- How do we estimate the parameters $\beta_0, \beta_1, \ldots, \beta_k$ and quantify the errors in the estimates?
- How can we tell whether a factor or variable X^i is relevant to the model? That is, is the coefficient β_i significantly different from 0?
- How do we obtain a forecast of Y or an interval estimate for $E(Y)$?
- What are some of the difficulties associated with this approach?

6.1.3.1 *The Linear Regression Model*

We start with the simplest model that has one independent variable and use it to illustrate the process of parameter estimation through to quantifying the errors in the forecast of the dependent variable.

A good grasp of the simple linear regression model allows us to handle multiple regression with ease as the basic concepts extend to models with more than one independent variable.

Suppose we are interested in the income of a worker. Without any information about the worker, our best guess that minimizes the mean squared error of the estimate is the average income of the working population. Can we improve our estimate by using some information about the worker, such as the number of years of formal education that the worker had? Consider the following data in Table 6.1 for 10 workers.

In the data set, the average number of years of formal education is 10.65 years and the average annual income is \$94,300. Without knowing a worker's educational background, the estimate for the annual income of a worker chosen from the data set (sample) will

Table 6.1: Sample dataset of no. of years of education against the annual income of a worker.

Datapoint	Years of education, X	Annual income, Y (thousands of \$)
1	7.0	0
2	9.5	129
3	15.0	150
4	10.5	107
5	9.5	81
6	18.5	181
7	3.5	26
8	11.5	86
9	16.5	105
10	5.0	78
	$\bar{X} = 10.65$	$\bar{Y} = 94.3$

be \$94,300. If one uses another estimate say \$$P$, then it is easy to check that $\sum_{i=1}^{n}(Y_i - 94{,}300)^2 \leq \sum_{i=1}^{n}(Y_i - P)^2$.

6.1.3.2 *Sample Correlation Coefficient*

We define the sample correlation and compute it from a set of data.

Linear correlation measures how closely the co-ordinate points formed from two variables are packed around a line.

To find out whether using the number of years of formal education X can help us improve our estimate of the annual income Y, we compute the correlation between the two quantities. To do that we first compute the sample variances of X and of Y as well as their covariance as follows:

$$\text{Sample variance of } X, \; s_x^2 = \frac{\sum_{i=1}^{n}(X_i - \bar{X})^2}{n-1}$$

where n is the sample size, $X_i = i$th observation and $\bar{X} = \frac{X_1 + \cdots + X_n}{n}$ is the sample mean of X.

$$\text{Sample variance of } Y, \; s_y^2 = \frac{\sum_{i=1}^{n}(Y_i - \bar{Y})^2}{n-1}$$

where n is the sample size, $Y_i = i$th observation and $\bar{Y} = \frac{Y_1 + \cdots + Y_n}{n}$ is the sample mean of Y.

$$\text{Sample covariance of } X \text{ and } Y, \; s_{xy} = \frac{\sum_{i=1}^{n}(X_i - \bar{X})(Y_i - \bar{Y})}{n-1}.$$

Table 6.2 summarizes the sample variance of X and Y and sample covariance of xy.

The sample covariance for the data set, $\sigma_{xy} = 208{,}510$ and this number is affected by the scales that are used to measure X and Y. We normalize the covariance by dividing it by the product of the standard deviations of X and Y to obtain the correlation between X and Y.

The sample correlation between X and Y is defined as

$$\rho = \frac{s_{xy}}{\sqrt{s_x^2 \times s_y^2}} = \frac{208{,}506}{\sqrt{23.84 \times 2{,}907{,}566{,}667}} = 0.7920$$

Aside: for two random variables A and B, the covariance of A and B is defined as

$$\text{covariance}(A,B) = E(AB) - E(A)E(B).$$

Table 6.2:   Sample variance of X and Y and sample covariance of xy.

Datapoint	Years of education, X	Annual income, Y (thousands of \$)	$(X_i - \bar{X})^2$	$(Y_i - \bar{Y})^2$ ($\times 1{,}000{,}000$)	$(X_i - \bar{X})(Y_i - \bar{Y})$ ($\times 1{,}000$)
1	7.0	0	13.32	8,892.49	344.20
2	9.5	129	1.32	1,204.09	−39.91
3	15.0	150	18.92	3,102.49	242.30
4	10.5	107	0.02	161.29	−1.91
5	9.5	81	1.32	176.89	15.30
6	18.5	181	61.62	7,516.89	680.60
7	3.5	26	51.12	4,664.89	488.35
8	11.5	86	0.72	68.89	−7.05
9	16.5	105	34.22	114.49	62.60
10	5.0	78	31.92	265.69	92.10
	$\bar{X} = 10.65$	$\bar{Y} = 94.3$	$s_x^2 = 23.84$	$s_y^2 = 2{,}907.57$	$s_{xy} = 208.51$

The correlation between A and B is defined as

$$\text{correlation}(A, B) = \frac{\text{covariance}(A, B)}{\sqrt{\sigma_A^2 \sigma_B^2}}$$

where σ_A^2 and σ_B^2 are the variances of A and B respectively.

Note that the definition is symmetric in that the correlation between X and Y is the same as the correlation between Y and X. The quantity ρ is known as the linear correlation coefficient or Pearson's correlation coefficient. Correlation coefficients are indicators of the strength of the relationship between two different variables. The correlation ρ can range from -1 to 1 (inclusive of the two values).

A positive correlation—when the correlation coefficient is greater than 0—signifies that both variables move in the same direction. When ρ is $+1$, it signifies that the two variables being compared have a perfect positive relationship; when one variable moves higher or lower, the other variable moves in the same direction with the same magnitude. A negative (inverse) correlation occurs when the correlation coefficient is less than 0. This is an indication that both variables move in the opposite direction. When ρ is -1, the relationship is said to be perfectly negatively correlated.

Finally, a value of zero indicates no relationship between the two variables that are being compared. The correlation of 0.792 in this example is moderately high. Those who are careful may wish to carry out a test of the hypothesis that allows them to conclude that ρ is significantly different from 0. A moderately high correlation means that a large amount of the variability in the annual income which ranges from \$0 to \$181,000 in the sample can be accounted for by considering the number of years of education the worker had. In other words, knowing the number of years of education a worker had, allows us to arrive at a better estimate of the worker's income. How do we use the number of years of education X to estimate the annual income Y?

6.1.3.3 *Scatterplot*

Visual inspection of the pictorial data can help us decide which approach or analysis may be appropriate and it may highlight potential issues that have to be addressed. Eye-balling a data set may allow us to identify features or relationships that could be further investigated or tested.

Scatterplots are used to analyze patterns of the relationship between two sets of continuous data. Scatterplots can visually show the strength of the relationship between the variables, form, direction, strength of association, or presence of outliers.

Before proceeding further, it is useful to take a look at the data in a scatterplot. Figure 6.1 illustrates the relationship between annual income and the number of education years.

We can see a positive correlation between the two variables. Note that it is customary to place the independent variable or predictor variable X on the horizontal axis and the dependent variable Y on the vertical axis. In this scenario, we may choose to believe that the number of years of education has a positive relation to the income of the workers. There may be several motivations for studying the relationship. An HR manager may be interested in using the level of education of a candidate to arrive at a salary to offer to the potential new hire. A researcher in the civil service may be interested to know how much income each additional year of education can bring for a worker to justify an increase in the funding for education.

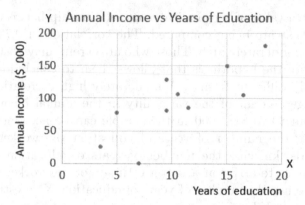

Figure 6.1:   Scatterplot diagram of annual income against number of years of education.

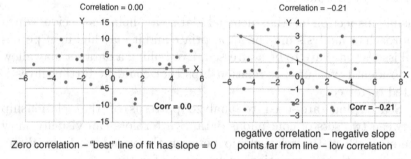

Figure 6.2:   Graphical illustration on zero correlation and negative correlation.

Figure 6.2 shows pictorially the difference between zero correlation and negative correlation. The chart on the left shows a plot of data points that where the X and Y variables have zero correlation. We see that the "best" line drawn to fit the data points has zero gradient.

The chart on the right shows a data set where the two variables are negatively correlated. We see that large values of X go with small or negative values of Y and conversely. The correlation at -0.21 is considered relatively low as the data points are far away from any line that one draws through them.

Figure 6.3 depicts two graphical illustrations on positive correlation. The chart on the left shows the plot of a data set where the two variables have a high positive correlation.

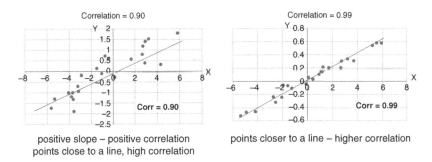

Figure 6.3: Graphical illustration on positive correlation.

The chart on the right shows a data set where the correlation is nearly perfect. Two variables are perfectly positively (resp. negatively) correlated if the correlation is 1 (resp. -1). Compared to the chart on the left, the points here are much more closely packed to the line. That is what correlation measures. How closely the data points are packed around the "best" line. The closer they are to a line with a positive (resp. negative) slope, the nearer the correlation will be to 1 (resp. -1). Note that the steepness of the line does not matter as long as the slope is not zero. If all the points lie on one line, the correlation will be perfect, that is 1 (resp. -1) if the slope is positive (resp. negative).

6.1.3.3.1 Non-linear Relationships and Outliers

Consider the scatterplots in Figure 6.4.

In Scatterplot 1, the correlation between X and Y is 0. Here is an example where the correlation is zero but we can discern some pattern or relationship between X and Y, namely that the points lie on a circle.

In Scatterplot 2, X and Y have a positive correlation of 0.61. Would you consider the relationship here to be a linear one? That is, is it reasonable to fit a single line to the points?

In Figure 6.5, we fitted a polynomial curve to the points. Does it look reasonable? Notice that there is a point in the lower left-hand corner that is far away from the regression curve. Without that point, the regression curve would probably go along the path indicated by the dotted curve.

Outliers can distort correlation computation leading us to believe that there is a strong correlation when none exists in reality or the

Scatterplot 1: correlation = 0.00

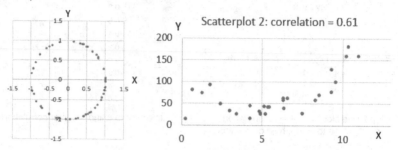

Figure 6.4: Scatterplot illustration on zero correlation and positive correlation.

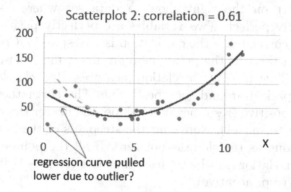

Figure 6.5: Polynomial regression curve for positive correlation scatterplot.

other way round. Outliers also present great difficulty for regression analysis as they can have a big impact on the slope of the regression line or curve. If the outliers are due to careless data collection or they occur once in a trillion (10^{12}) observations on average, then they may be ignored. But if the outliers are a feature of the distribution then we should keep them. In a nutshell, we should look at them seriously before deciding one way or the other.

6.1.3.4 *Least-squares Regression Line*

We now describe why the least-squares regression line is chosen to minimize the sum of the squares of the residuals rather than to minimize other sums of squared distances such as the perpendicular distance to the line.

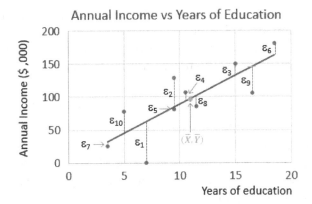

Figure 6.6: Least-square regression line.

The least-squares regression line is designed solely to obtain a good prediction of the dependent variable.

Going back to the income and education data set, a linear relationship between the two variables seems reasonable. Which linear relationship or line should we use? As minimizing the squared error is a common criterion for comparing estimators, it is usual to fit a line to minimize the sum of the squares of the vertical distances of the points to the line as depicted in the graphical illustration in Figure 6.6.

The regression line is chosen so that the sum of the squares of the errors $\sum_{i=1}^{n} \varepsilon_i^2$ is a minimum. This is known as the least-squares regression line. Note that the point $(\bar{X}, \bar{Y})$ formed by the sample means of X and Y lies on this line. The least-squares regression line has the property that the errors ε_i sum to zero, that is $\sum_{i=1}^{n} \varepsilon_i = 0$.

The equation of the regression line is $Y = \hat{\beta}_0 + \hat{\beta}_1 X$ where $\hat{\beta}_1 = \frac{s_{xy}}{s_x^2} = \frac{208,506}{23.84} = 8,747.5$ and

$$\hat{\beta}_0 = \bar{Y} - \hat{\beta}_1 \bar{X} = 94,300 - 8,747.47 \times 10.65 = 1,139.5.$$

Note that $\hat{\beta}_1$ is the least-squares estimate of the slope of the regression line and $\hat{\beta}_0$ is the least-squares estimate of the intercept.

With the least-squares regression line, $Y = 1,139.5 + 8,747.5X$, a HR manager estimates that a worker with 12 years of education has an annual income of

$$\hat{Y} = 1,139.5 + 8,747.5 \times 12 = \$106,109.50 \text{ on average.}$$

Using the regression equation, a policy analyst may conclude that each additional year of education increases the annual income of a worker by \$8,747.50 on average. Notice that the forecasts are average values since many of the points do not lie on the line.

6.1.3.5 *Coefficient of Determination R^2 and Correlation*

We would like to understand why the coefficient of determination is a key indicator of the usefulness of a linear model and its relation to linear correlation. An important result is that the higher the coefficient of determination, the lower will be the error in the prediction of the dependent variable.

How much improvement do we get if we use the regression line to forecast annual income? If we don't use X, then our estimate for Y will be the sample mean, $\bar{Y} = \$94,300$. The total sum of squares (TSS) measures the total variation in the dependent variable Y and it is defined as the sum of the squares of the differences between Y_i and the sample mean $\bar{Y}$, $\text{TSS} = \sum_{i=1}^{n}(Y_i - \bar{Y})^2$. Figure 6.7 graphs the relation of the sum of squared errors, the TSS, and the residual sum of squares.

The regression sum of squares, $\text{RSS} = \sum_{i=1}^{n}(\hat{Y}_i - \bar{Y})^2$ is the amount of variation in Y that is explained by the regression model. In some books, this is written as SSR for the sum of squares due to regression. This quantity is also called the model sum of squares (MSS) or explained sum of squares (ESS). The larger RSS is, the

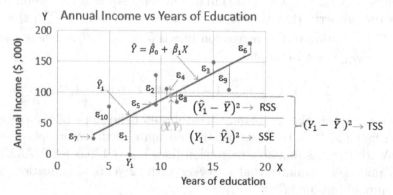

Figure 6.7: Illustration of the sum of squared errors, the TSS, and the residual sum of squares.

larger the amount of variation in Y that the model can account for or explain.

The sum of squared errors (SSE) $= \sum_{i=1}^{n}(Y_i - \hat{Y}_i)^2 = \sum_{i=1}^{n}\varepsilon_i^2$ is the amount of variation in the dependent variable Y that is not explained by the regression model. SSE is computed by summing the squares of the residuals $(Y_i - \hat{Y}_i)$. The residuals are the errors between the actual observation Y_i and the predicted value $\hat{Y}_i$ using the regression line. Sometimes this quantity is called the residual sum of squares and given the abbreviation RSS, which we use here for the RSS. Please check carefully the definitions of the abbreviations such as TSS, SST, RSS, MSS, ESS, and SSE when you read different books on regression.

Note that $Y_i - \bar{Y} = (\hat{Y}_i - \bar{Y}) + (Y_i - \hat{Y}_i)$ (*). It turns out that squaring both sides of equation (*) and summing over i we have

$$\sum_{i=1}^{n}(Y_i - \bar{Y})^2 = \sum_{i=1}^{n}(\hat{Y}_i - \bar{Y})^2 + \sum_{i=1}^{n}(Y_i - \hat{Y}_i)^2.$$

This last equation does not follow from equation (*) in a straightforward manner and has to be checked formally. Using the result, TSS can be written as the sum of the two components RSS and SSE as indicated in the diagram.

Total variation, TSS = explained variation + unexplained variation

$$= \text{RSS} + \text{SSE}.$$

Let us consider the values in the data set. We use the regression line to estimate Y for those values of X in the data set as shown in Table 6.3.

From Table 6.3, we see that the total variation in Y, TSS $= 26{,}168.10 \times 10^6$ and the sum of squared errors SSE $= 9{,}753.04 \times 10^6$. This means that the amount of variation in Y that is explained by the model is RSS $=$ TSS $-$ SSE $= 26{,}168.10 \times 10^6 - 9{,}753.04 \times 10^6 = 16{,}415.06 \times 10^6$.

The proportion of the total variation in Y that is explained by the regression model is known as the coefficient of determinant and is usually denoted as R^2. In the example,

Coefficient of determination, $R^2 = \frac{\text{TSS}-\text{SSE}}{\text{TSS}} = \frac{16{,}415.06 \times 1{,}000{,}000}{26{,}168.10 \times 1{,}000{,}000} = 0.6273$.

Table 6.3: Total sum of square and sum of squared error.

Data point	Years of education, X	Annual income, Y (thousands of \$)	Estimate $= \bar{Y}$ $(Y_i - \bar{Y})^2$ ($\times 1{,}000{,}000$)	$\hat{Y}_i = \hat{\beta}_0 + \hat{\beta}_1 X$ (thousands of \$)	Estimate $= \hat{Y}_i$ $(Y_i - \hat{Y}_i)^2$ ($\times 1{,}000{,}000$)
1	7.0	0	8,892.49	62.37	3,890.24
2	9.5	129	1,204.09	84.24	2,003.42
3	15.0	150	3,102.49	132.35	311.47
4	10.5	107	161.29	92.99	196.34
5	9.5	81	176.89	84.24	10.50
6	18.5	181	7,516.89	162.97	325.17
7	3.5	26	4,664.89	31.76	33.13
8	11.5	86	68.89	101.74	247.60
9	16.5	105	114.49	145.47	1,638.04
10	5.0	78	265.69	44.88	1,097.14
	$\bar{X} = 10.65$	$\bar{Y} = 94.3$	TSS $= 26{,}168.10$	Average $= 94.3$	SSE $= \sum_{i-1}^{n} \varepsilon_i^2 = 9{,}753.04$

The regression model can account for 62.73% of the total variation in Y. If $R^2 = 0.6273$, what is the value of R? $R = \sqrt{0.6273} = 0.7920$. Does this number look familiar? Yes, $R = 0.7920$ is the sample correlation coefficient between X and Y that we computed earlier. The higher the correlation between X and Y, the higher the proportion of the total variation in Y that can be explained by the regression model.

6.1.3.6 *Adjusted* R^2

We next explain the rationale for defining an adjusted R^2 and how it is computed. A key rationale for the use of an adjusted R is the more factors or independent variables there are, the greater will be the mark-down of the coefficient of determination.

In general, as more independent variables (which are not 100% perfectly correlated to existing ones) are added to a linear equation model, the value of R^2 will go up even though these new variables may be irrelevant and are not useful for explaining or forecasting the dependent variable. For example in Table 6.4 below, we add a variable V, which takes 1 or -1 at random to the income-education data set.

We consider the regression model $Y = \beta_0 + \beta_1 X + \beta_2 V + \varepsilon$.

Table 6.4: Spurious variable and adjusted R^2.

Data point	Years of education, X	Spurious variable, V	Annual income, Y (thousands of $)
1	7.0	1	0
2	9.5	-1	129
3	15.0	1	150
4	10.5	-1	107
5	9.5	1	81
6	18.5	-1	181
7	3.5	-1	26
8	11.5	-1	86
9	16.5	1	105
10	5.0	1	78
	$\bar{X} = 10.65$	$\bar{V} = 0$	$\bar{Y} = 94.3$

Table 6.5: R^2 and adjusted R^2.

Number of independent variables	R^2	Adjusted R^2
One, X	0.6273	0.5807
Two, X and V	0.6741	0.5809

The least-squares estimates for the parameters are $\hat{\beta}_0 = 1.41$, $\hat{\beta}_1 = 8.72$, $\hat{\beta}_2 = -11.06$.

The R^2 for this model is 0.6741 which is higher than that for the regression with one variable X. Recall that the model with X as the single predictor variable has $R^2 = 0.6273$. The R^2 quantity goes up even though the new variable V is a random quantity that has nothing to do with annual income. To penalize the inclusion of more variables, an adjusted R^2 quantity is computed as follows:

$$\text{Adjusted } R^2, \ R^2_{\text{adj}} = 1 - \frac{n-1}{n-k-1}(1 - R^2),$$

where n is the number of observations or data points and k is the number of independent variables excluding the constant β_0.

For the model with one independent variable X,

$$\text{adjusted } R^2, \ R^2_{\text{adj}} = 1 - \frac{10-1}{10-1-1}(1 - R^2) = 0.5807.$$

For the model with two independent variables, X and V,

$$\text{adjusted } R^2, \ R^2_{\text{adj}} = 1 - \frac{10-1}{10-2-1}(1 - R^2) = 0.5809.$$

Table 6.5 depicts the computed R^2 and adjusted R^2.

In this example, we see that the adjustment did not help much as the bigger model with the irrelevant independent variable V still has a higher adjusted R^2.

6.1.3.7 Assumptions of a Linear Model with One Independent Variable

It is important to know the assumptions that underlie the model to anticipate, as well as to detect, likely violations of the assumptions in practice. The main assumptions are about the validity of a linear relationship, the nature of the errors, and their relationship to the independent variables.

In a linear model with one independent variable X, the dependent variable Y satisfies a linear relationship that may be written as

$$Y = \beta_0 + \beta_1 X + \varepsilon$$

where β_0 and β_1 are constants and ε is an error term.

The parameter β_1 is the slope of the regression line and β_0 is the intercept of the line.

The following assumptions are typically adopted for the analysis of the model.

- The dependent variable Y is related to the independent variable X by the linear relationship, where the parameters β_0 and β_1 are assumed to be constants that are fixed.
- The independent variable X takes on two or more values in the sample of observations. This is to avoid the problem of multi-collinearity. This assumption is known as the identification condition.
- For each value of X, the error term ε has (conditional) mean zero and a constant (conditional) variance σ^2. That is, $E(\varepsilon|X) = 0$ and variance$(\varepsilon|X) = \sigma^2$.
- Across observations, the error terms are uncorrelated. That is, the correlation between ε_i and ε_j is 0 for $i \neq j$. This assumption may be strengthened by assuming that the error terms are independent.
- The values for the independent variable X are assumed to be fixed for repeated samples. However, the analysis can still be carried out when X is a random variable. Whether X is random or fixed, the way that the values of X are generated is not related to ε.
- The error term has the normal distribution, $\varepsilon \sim N(0, \sigma^2)$. This assumption is not necessary for the estimation of the parameters or analysis of the linear model. It is adopted to make testing hypotheses easier when the sample size is small.

6.1.3.8 *Confidence Interval Estimate for the Parameters* β_0 *and* β_1

Key information from regression models is point and interval estimates for the population parameters. The goal of the model is to predict or forecast the dependent variable Y for a given value of X. The parameters are estimated so that the sum of squared errors in the forecast based on the data set is a minimum. These are known as the least-squares estimators.

For a given linear model, $Y = \beta_0 + \beta_1 X + \varepsilon$, we usually estimate the parameters β_0 and β_1 from a sample of data points using the criterion of minimizing the sum of squares of the residuals or errors. These estimates are called the least-square estimates and they have the following two properties.

1. The least-squares estimators $\hat{\beta}_0$ and $\hat{\beta}_1$ are linear estimators in that they both can be expressed as linear combinations of the values of Y. In particular, we can write $\hat{\beta}_0$ and $\hat{\beta}_1$ as follows:

$$\hat{\beta}_1 = \frac{s_{xy}}{s_x^2} = \sum_{i=1}^{n} \left(\frac{X_i - \bar{X}}{\sum_{i=1}^{n}(X_i - \bar{X})^2} \right) Y_i \quad \text{and}$$

$$\hat{\beta}_0 = \bar{Y} - \hat{\beta}_1 \bar{X} = \sum_{i=1}^{n} \left[\frac{1}{n} - \left(\frac{X_i - \bar{X}}{\sum_{i=1}^{n}(X_i - \bar{X})^2} \right) \bar{X} \right] Y_i.$$

2. Both $\hat{\beta}_0$ and $\hat{\beta}_1$ are unbiased estimators. That is, $E(\hat{\beta}_0) = \beta_0$ and $E(\hat{\beta}_1) = \beta_1$.

The least-square estimators $\hat{\beta}_0$ and $\hat{\beta}_1$ are the best linear unbiased estimators (filled circle) in that they have the smallest mean squared error among all unbiased linear estimators. Notice that we are comparing unbiased estimators. In some cases, some biased estimators can have a lower mean squared error. If $\hat{\theta}$ is an estimator for a parameter θ, the mean square error $\text{MSE}(\hat{\theta})$ may be written as $\text{MSE}(\hat{\theta}) = E(\hat{\theta} - \theta)^2 = E(\hat{\theta} - E(\hat{\theta}))^2 + (E(\hat{\theta}) - \theta)^2 = \text{variance}(\hat{\theta}) + (\text{bias})^2$.

A biased estimator can have a lower MSE compared to an unbiased estimator if its variance is small and its bias is not too big.

For the income-education data set, the least-squares estimates for the parameters are $\hat{\beta}_0 = 1{,}139.5$ and $\hat{\beta}_1 = 8{,}747.5$. These are point estimates and using another sample, we will obtain different estimates as illustrated in Figure 6.8.

To obtain a CI estimate for these parameters, let us assume that the errors $\varepsilon_i \sim N(0, \sigma^2)$ are independent and follow a normal distribution with mean zero and variance σ^2. Under the assumption for

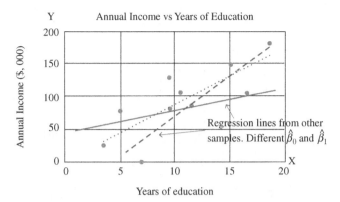

Figure 6.8: Different regression line for no. of years of education against annual income of a worker.

the error terms, it can be shown that the variances of the estimators are

$$\text{variance}(\hat{\beta}_0) = \left[\frac{1}{n} + \frac{(\bar{X})^2}{\sum_{i=1}^n (X_i - \bar{X})^2} \right] \sigma^2$$

and

$$\text{variance}(\hat{\beta}_1) = \frac{\sigma^2}{\sum_{i=1}^n (X_i - \bar{X})^2}.$$

If we know σ^2 then it is easy to compute the variances of the estimators using the above equations. Since we usually do not know σ^2 we make use of the fact that

$$\frac{\sum_{i=1}^n \varepsilon_i^2}{\sigma^2} \sim \chi^2(n-2) \text{ to arrive at the estimator } \hat{\sigma}^2 = \frac{\sum_{i=1}^n \varepsilon_i^2}{n-2} \text{ for } \sigma^2.$$

Note that $\hat{\sigma}^2$ is dependent on $\hat{\beta}_0$ and $\hat{\beta}_1$ as the residuals ε_i are computed based on the regression line $Y = \hat{\beta}_0 + \hat{\beta}_1 X$. The estimator $\hat{\sigma}^2$ is an example of a model-dependent estimate. The quantity $\text{SSE} = \sum_{i-1}^n \varepsilon_i^2$ has $n - 2$ degrees of freedom as we estimated two parameters β_0 and β_1. The estimator $\hat{\sigma}^2$ is unbiased. That is $E(\hat{\sigma}^2) = \sigma^2$.

We estimate the standard deviation or standard error of $\hat{\beta}_0$ and $\hat{\beta}_1$ as

$$\text{se}(\hat{\beta}_0) = \sqrt{\left[\frac{1}{n} + \frac{(\bar{X})^2}{\sum_{i=1}^{n}(X_i - \bar{X})^2}\right]\hat{\sigma}^2}$$

$$= \sqrt{\left[\frac{1}{n} + \frac{(\bar{X})^2}{\sum_{i=1}^{n}(X_i - \bar{X})^2}\right]\frac{\sum_{i=1}^{n}\varepsilon_i^2}{(n-2)}} \quad \text{and}$$

$$\text{se}(\hat{\beta}_1) = \sqrt{\frac{\hat{\sigma}^2}{\sum_{i=1}^{n}(X_i - \bar{X})^2}} = \sqrt{\frac{\sum_{i=1}^{n}\varepsilon_i^2}{(n-2)\sum_{i=1}^{n}(X_i - \bar{X})^2}}.$$

Let us compute the confidence intervals for the parameters β_0 and β_1 distinguishing between two cases where the variance σ^2 is known and where it is unknown.

Case 1: The error term $\varepsilon \sim N(0, \sigma^2)$ is normally distributed and the variance σ^2 is known.

It follows that $\hat{\beta}_1 \sim N\left(\beta_1, \frac{\sigma^2}{\sum_{i=1}^{n}(X_i - \bar{X})^2}\right)$ and

$$\hat{\beta}_0 \sim N\left(\beta_0, \left[\frac{1}{n} + \frac{(\bar{X})^2}{\sum_{i=1}^{n}(X_i - \bar{X})^2}\right]\sigma^2\right)$$

A $(1 - \alpha)$ CI for the parameter β_0 is

$$\left(\hat{\beta}_0 - Z_{\alpha/2}\sqrt{\left[\frac{1}{n} + \frac{(\bar{X})^2}{\sum_{i=1}^{n}(X_i - \bar{X})^2}\right]\sigma^2},\right.$$

$$\left.\hat{\beta}_0 + Z_{\alpha/2}\sqrt{\left[\frac{1}{n} + \frac{(\bar{X})^2}{\sum_{i=1}^{n}(X_i - \bar{X})^2}\right]\sigma^2}\right).$$

Recall that for a standard normal random variable $Z \sim N(0, 1)$, we have $P(Z < |Z_{\alpha/2}|) = 1 - \alpha$.

A $(1 - \alpha)$ CI for the parameter β_1 is

$$\left(\hat{\beta}_1 - Z_{\alpha/2}\sqrt{\frac{\sigma^2}{\sum_{i=1}^{n}(X_i - \bar{X})^2}}, \hat{\beta}_1 + Z_{\alpha/2}\sqrt{\frac{\sigma^2}{\sum_{i=1}^{n}(X_i - \bar{X})^2}}\right).$$

Case 2: The error term $\varepsilon \sim N(0, \sigma^2)$ is normally distributed and the variance σ^2 is unknown.

We estimate the variance σ^2 using the estimator $\hat{\sigma}^2 = \frac{\sum_{i=1}^n \hat{\varepsilon}_i^2}{n-2}$. As we estimated the variance, the relevant distribution for the parameters will be the t-distribution with $n-2$ degrees of freedom.

A $(1-\alpha)$ level CI for the parameter β_0 is

$$\left(\hat{\beta}_0 - t_{n-2,\alpha/2}\sqrt{\left[\frac{1}{n} + \frac{(\bar{X})^2}{\sum_{i=1}^n (X_i - \bar{X})^2}\right]\hat{\sigma}^2},\right.$$

$$\left.\hat{\beta}_0 + t_{n-2,\alpha/2}\sqrt{\left[\frac{1}{n} + \frac{(\bar{X})^2}{\sum_{i=1}^n (X_i - \bar{X})^2}\right]\hat{\sigma}^2}\right),$$

where for a random variable $T \sim t_{n-2}$ having the t-distribution with $n-2$ degrees of freedom, we have $P(T < |t_{n-2,\alpha/2}|) = 1 - \alpha$.

A $(1-\alpha)$ level CI for the parameter β_1 is

$$\left(\hat{\beta}_1 - t_{n-2,\alpha/2}\sqrt{\frac{\hat{\sigma}^2}{\sum_{i=1}^n (X_i - \bar{X})^2}}, \hat{\beta}_1 + t_{n-2,\alpha/2}\sqrt{\frac{\hat{\sigma}^2}{\sum_{i=1}^n (X_i - \bar{X})^2}}\right).$$

For the income-education data set, let us derive the 95% CI for the parameters β_0 and β_1. We have

$$\hat{\sigma}^2 = \frac{\sum_{i=1}^n \hat{\varepsilon}_i^2}{n-2} = 1{,}219.13 \times 10^6, \sum_{i=1}^n (X_i - \bar{X})^2 = 214.53, \bar{X} = 10.65,$$

$\alpha = 0.05$, and $t_{8,0.025} = 2.3060$.

Using the above numbers, a 95% CI for the parameter β_0 is $(-62{,}703, 64{,}982)$.

A 95% CI for the parameter β_1 is $(3{,}250, 14{,}245)$.

What conclusions can we draw from these confidence intervals?

A question of interest is whether, or not, the number of years of education has any impact on income. That is, is the slope β_1 in the model significantly different from 0?

The testing hypothesis is equivalent to constructing confidence intervals. The symmetric 95% CI for β_1 is $(3{,}250, 14{,}245)$ and it does not contain the value 0. We can conclude that a two-tail test of the null hypothesis $\beta_1 = 0$ at the 5% level of significance will reject the null hypothesis in favor of the conclusion that β_1 is significantly different from 0.

Table 6.6: Properties of least-square estimators.

Parameter	Intercept parameter β_0	Slope parameter β_1
Least-square estimator	$\hat{\beta}_0 = \bar{Y} - \left(\dfrac{\sum_{i=1}^{n}(X_i - \bar{X})Y_i}{\sum_{i=1}^{n}(X_i - \bar{X})^2} \right) \bar{X}$	$\hat{\beta}_1 = \dfrac{\sum_{i=1}^{n}(X_i - \bar{X})Y_i}{\sum_{i=1}^{n}(X_i - \bar{X})^2}$
Mean	$E(\hat{\beta}_0) = \beta_0$ unbiased estimator	$E(\hat{\beta}_1) = \beta_1$ unbiased estimator
Variance	$\text{var}(\hat{\beta}_0) = \left[\dfrac{1}{n} + \dfrac{(\bar{X})^2}{\sum_{i=1}^{n}(X_i - \bar{X})^2} \right] \sigma^2$	$\text{var}(\hat{\beta}_1) = \dfrac{\sigma^2}{\sum_{i=1}^{n}(X_i - \bar{X})^2}$
Test statistic when σ^2 is estimated using $\hat{\sigma}^2$	$T_0 = \dfrac{\hat{\beta}_0 - \beta_0}{\sqrt{\left[\frac{1}{n} + \frac{(\bar{X})^2}{\sum_{i=1}^{n}(X_i - \bar{X})^2} \right] \hat{\sigma}^2}}$ $\sim t_{n-2}$, t-distribution with $n-2$ degrees of freedom	$T_1 = \dfrac{\hat{\beta}_1 - \beta_1}{\sqrt{\frac{\hat{\sigma}^2}{\sum_{i=1}^{n}(X_i - \bar{X})^2}}}$ $\sim t_{n-2}$, t-distribution with $n-2$ degrees of freedom

What about β_0? Could $\beta_0 = 0$ be consistent with the data set? If we do a two-tail test of the null hypothesis $\beta_0 = 0$ at the 5% level of significance, we will find that the null hypothesis is not rejected since the value 0 is found in the symmetric 95% CI $(-62{,}703.30, 64{,}982.30)$.

Notice that the confidence intervals that we constructed for the parameters β_0 and β_1 both have 95% confidence and the interval for β_0 is more than 11 times as long as that for β_1. This says that there is greater uncertainty in estimating β_0.

Table 6.6 summarizes the properties of the least-square estimators.

6.1.3.9 *Confidence Interval Estimate for the Value of E(Y)*

For a given value of the independent variable X, we would like to obtain interval estimates for the mean value $E(Y|X)$ of Y given X, and for a single observation Y. The key to understanding the difference between the two-interval estimates is to realize that the variance of $E(\hat{Y}|X)$ depends on how far X is away from the sample mean $\bar{X}$.

Recall that the independent variable X and the dependent variable Y satisfy a linear relationship that may be written as:

$$Y = \beta_0 + \beta_1 X + \varepsilon$$

where β_0 and β_1 are fixed constants and ε is an error term with variance $\text{var}(\varepsilon) = \sigma^2$.

For a given value x^* of the independent variable, we wish to estimate $E(Y|X = x^*)$. The quantity $E(Y|X = x^*)$ is the expected value of Y given $X = x^*$. This is the average value of Y when the independent variable $X = x^*$. Writing $\theta = E(Y|X = x^*)$, we use the regression line to obtain an estimate $\hat{\theta} = \hat{\beta}_0 + \hat{\beta}_1 x^*$ for θ. This is a point estimate of θ. We expect that there will be error in the estimate $\hat{\theta}$ as we estimated the population parameters β_0 and β_1 using the least-square estimators. To construct a CI for θ we need an estimate of the standard error or standard deviation of $\hat{\theta}$. It turns out that the standard error or deviation of $\hat{\theta}$ is given by

$$\text{se}(\hat{\theta}) = \sigma \sqrt{\frac{1}{n} + \frac{(x^* - \bar{X})^2}{\sum_{i=1}^{n}(X_i - \bar{X})^2}}.$$

Notice that $\text{se}(\hat{\theta})$ depends on the location of x^* relative to the mean $\bar{X}$. In particular when $x^* = \bar{X}$, we have $\text{se}(\hat{\theta}) = \sigma/\sigma\sqrt{n}$. The closer x^* is to the mean $\bar{X}$, the smaller $\text{se}(\hat{\theta})$ will be. The idea is that the estimate $E(Y|X = x^*)$ is more stable when x^* is near the mean $\bar{X}$, and less reliable when x^* is far away from $\bar{X}$. Why is this so?

Let μ_X and μ_Y denote the population means of X and Y respectively. For a sample, the point $(\bar{X}, \bar{Y})$ formed by the sample means will be quite close to (μ_X, μ_Y) when the sample size is reasonably large. Each regression line obtained using the least-squares criterion will always pass through the point $(\bar{X}, \bar{Y})$. Thus almost all the least-squares regression lines will pass near the point (μ_X, μ_X). So the estimation for $E(Y)$ when x^* is near $\bar{X}$ will be more reliable compared to the case when x^* is far away from $\bar{X}$. If x^* is far away from $\bar{X}$ any small error in the least-square estimate of the slope $\hat{\beta}_1$ will be magnified. The situation is illustrated in Figure 6.9.

For the income-education data, let us construct a 95% CI for $E(Y)$ when $X = 12$ years. As before, we estimate σ^2 using $\hat{\sigma}^2 = \frac{\sum_{i=1}^{n} \varepsilon_i^2}{n-2} =$

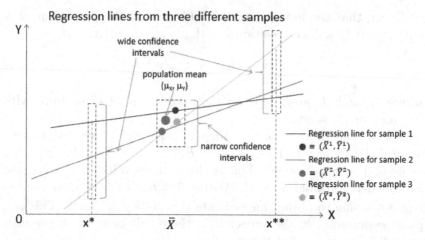

Figure 6.9: Regression line of three different samples.

$1,219.13 \times 10^6$ and

$$\hat{\sigma}\sqrt{\frac{1}{n} + \frac{(x^* - \bar{X})^2}{\sum_{i=1}^{n}(X_i - \bar{X})^2}}$$

$$= \sqrt{1,219.13 \times 10^6} \times \sqrt{\frac{1}{10} + \frac{(12 - 10.65)^2}{9 \times 23.84}}$$

$$= 30,997.44.$$

The distribution of $T = \dfrac{\hat{\theta} - \theta}{\hat{\sigma}\sqrt{\frac{1}{n} + \frac{(x^* - \bar{X})^2}{\sum_{i=1}^{n}(X_i - \bar{X})^2}}}$ is the t-distribution with $n - 2$ degrees of freedom.

The critical level is $t_{8,0.025} = 2.3060$.

A 95% CI for $E(Y|X = 12)$ is $(34,629, 177,590)$.

6.1.3.10 *Confidence or Prediction Interval for an Observation Y*

In the previous section, we obtain a CI for the expected value of Y, $E(Y|X = x^*)$ for a given value of $X = x^*$. What if we wish to have a CI for a single observation Y given $X = x^*$? A single observation Y is more variable compared to the mean $E(Y)$ as illustrated in Figure 6.10.

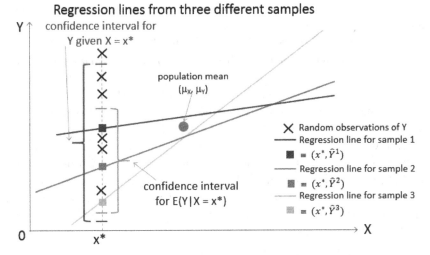

Figure 6.10: Illustration on confidence intervals for Y.

It turns out that the variance of Y given $X = x^*$ is greater by one copy of $\text{var}(\varepsilon) = \sigma^2$, namely

$$\text{variance}(\hat{Y}) = \sigma^2 \left(1 + \frac{1}{n} + \frac{(x^* - \bar{X})^2}{\sum_{i=1}^{n}(X_i - \bar{X})^2} \right).$$

When the value of σ^2 is unknown, we estimate it using $\hat{\sigma}^2 = \frac{\sum_{i=1}^{n} \varepsilon_i^2}{n-2}$. The limits of a 95% CI for Y are

$$(\hat{\beta}_0 + \hat{\beta}_1 x^*) \pm t_{n-2,0.025}\hat{\sigma}\sqrt{1 + \frac{1}{n} + \frac{(x^* - \bar{X})^2}{\sum_{i=1}^{n}(X_i - \bar{X})^2}}.$$

With the income-education data, for $x^* = 12$, a 95% CI for Y is (21,338, 190,881). This interval is sometimes known as a prediction interval.

6.1.3.11 *Test of Hypothesis for the Parameters*

We would like to find out if the slope and intercept parameters are significantly different from 0. We can conclude that a parameter is significantly different from 0 if the least-squares estimate for it is sufficiently far away from 0.

From a sample of observations, we form the least-squares estimates $\hat{\beta}_0$ and $\hat{\beta}_1$ of the population parameters. For the income-education data, we have $\hat{\beta}_0 = 1{,}139.5$ and $\hat{\beta}_1 = 8{,}747.5$.

It is natural to wonder whether the independent variable X is relevant for forecasting the annual income. That is, could the estimate $\hat{\beta}_1 = 8{,}747.5$ be consistent with the condition that the true value of the slope $\beta_1 = 0$? When the true value is $\beta_1 = 0$, we could still obtain $\hat{\beta}_1 = 8{,}747.5$ by random chance or fluctuations. How likely is it that we will observe such a sample? Note that when $\beta_1 = 0$, the regression line is flat and the independent variable is uncorrelated with the dependent variable. To find out whether $\beta_1 = 0$ or not, we conduct a two-tail test of the hypothesis that $\beta_1 = 0$ at the 5% level of significance.

The null hypothesis, H_0: $\beta_1 = 0$.

The alternative hypothesis, H_1: $\beta_1 \neq 0$.

Level of significance: 5%.

Test statistic, $T_1 = \dfrac{\hat{\beta}_1 - \beta_1}{\sqrt{\dfrac{\hat{\sigma}^2}{\sum_{i=1}^{n}(X_i - \bar{X})^2}}} = \dfrac{8{,}747.5 - 0}{\sqrt{\dfrac{1{,}219.13 \times 1{,}000{,}000}{214.53}}} = 3.6694.$

Critical value: $T_{8,0.025} = 2.3060$.

Critical or rejection region: $T_1 < -2.3060$ or $T_1 > 2.3060$.

Conclusion: since $T_1 = 3.6694 > 2.3060$, we reject the null hypothesis.

Thus the slope is significantly different from 0 at the 5% level of significance.

The p-value is 0.6315%. That is if the null hypothesis is true, then the chance of us seeing the given data or other data sets more adverse against the null hypothesis is 0.6315%. As this is a very rare event if the null hypothesis is true, it is more reasonable to accept the conclusion that the null hypothesis is incorrect.

We already made this conclusion earlier when we derive the 95% CI for β_1. The symmetric 95% CI for β_1 is $(3{,}250.24, 14{,}244.76)$ and since 0 is not in the interval, we concluded that the null hypothesis $\beta_1 = 0$ will be rejected at the 5% level of significance.

We now consider whether the intercept β_0 is significantly different from 0. From the data, the least-square estimate of this parameter is $\hat{\beta}_0 = 1{,}139.5$. We perform a two-tail test of the hypothesis $\beta_0 = 0$ at the 5% level of significance.

Null hypothesis, H_0: $\beta_0 = 0$.

Alternative hypothesis, H_1: $\beta_0 \neq 0$.

Level of significance: 5%.

Test statistic, $T_0 = \dfrac{\hat{\beta}_0 - \beta_0}{\sqrt{\left[\frac{1}{n} + \frac{(\bar{X})^2}{\sum_{i=1}^{n}(X_i - \bar{X})^2}\right]\hat{\sigma}^2}} = 0.0412.$

Critical value: $T_{8,0.025} = 2.3060$.

Critical or rejection region: $T_1 < -2.3060$ or $T_1 > 2.3060$.

Conclusion: since $T_0 = 0.0412 \in (-2.3060,\ 2.3060)$, we do not reject the null hypothesis.

This is a conclusion we made as to the symmetric 95% CI for β_0, $(-62{,}703,\ 64{,}982)$ contains the value 0.

6.1.4 *Linear Models and ANOVA*

Linear models may be analyzed by comparing various sums of squared deviations, which is a technique in ANOVA. In the case of a simple linear model, testing whether the independent variable plays a significant role in the model is the same as performing a test of hypothesis in the ANOVA framework.

ANOVA stands for analysis of variance and it is a tool for investigating whether the average value of a quantity among various groups are the same. Suppose we classify a worker according to his or her education background as "primary level", "secondary level" and "tertiary level". ANOVA can help us answer the question: do the workers in these three groups have the same average salary? It does so by considering a test statistic formed from the ratio of the sum of squared deviations. As the sum of squared deviations is related to variance, hence the name.

Consider the income-education data set. Suppose we believe that the years of education is not relevant to the salary of a worker, we may simply model the salary as

Model A: $Y = \beta_0 + \varepsilon$ where β_0 is a constant and $\varepsilon \sim N(0, \sigma^2)$ is an error term.

What can β_0 be? Notice that $E(Y) = \beta_0 + E(\varepsilon) = \beta_0$. That is, $\beta_0 = E(Y) = \mu_Y$ the population mean of Y. For a given sample of observations, this model will always estimate Y using the sample mean $\bar{Y}$. That is, $\hat{\beta}_0 = \bar{Y}$ and the predicted value $\hat{Y} = \hat{\beta}_0 = \bar{Y}$.

The sum of squared errors, $\mathrm{SS}_A = \mathrm{TSS} = \sum_{i=1}^{10}(\hat{Y} - Y_i)^2 = \sum_{i=1}^{10}(\bar{Y} - Y_i)^2 = 2.6168 \times 10^{10}$.

The degree of freedom is $\mathrm{df}_A = n - 1 = 9$ as we estimated the population mean.

The simple linear model using the years of education as an independent variable is given below.

Model B: $Y = \beta_0 + \beta_1 X + \varepsilon$ where β_0 and β_1 are constants and $\varepsilon \sim N(0, \sigma^2)$ is an error term.

Model B is the linear regression model that we studied earlier. Using the data set, we estimate the parameters using the least-squares estimates $\hat{\beta}_0$ and $\hat{\beta}_1$. For a given X_i, the model estimates the value of Y using the regression line, namely $\hat{Y}_i = \hat{\beta}_0 + \hat{\beta}_1 X_i$.

In this case the sum of the squared errors, $SS_B = SSE = \sum_{i=1}^{n} (\hat{Y}_i - Y_i)^2 = \sum_{i=1}^{n} \varepsilon_i^2 = 9.7530 \times 10^9$.

The degree of freedom is $df_B = n - 2 = 8$ as we estimated the two parameters.

We see that Model B is a bigger model that includes model A as a special case when $\beta_1 = 0$.

To find out whether β_1 is significantly different from 0, we can perform the following test of hypothesis.

Null hypothesis H_0: $\beta_1 = 0$.

Alternative hypothesis H_1: $\beta_1 \neq 0$.

Level of significance: 5%.

The test statistics are formed by the ratio of the sum of squares divided by their degrees of freedom.

$$F = \frac{(SS_A - SS_B)/(df_A - df_B)}{SS_B/df_B} = \frac{(TSS - SSE)/(n - 1 - (n - 2))}{SSE/(n - 2)}$$

$$= \frac{RSS/1}{SSE/(n - 2)}.$$

Under the null hypothesis $F \sim F_{1,n-2}$ has the F distribution with 1 and $n - 2$ degrees of freedom.

We reject the null hypothesis for large values of F.

Critical region: $F > F_{1,n-2}(0.05) = F_{1,8}(0.05) = 5.3177$.

From the data, we have $F = \frac{1.6415 \times 10^{10}}{9.7530 \times 10^9/8} = 13.4646 > F_{1,8}(0.05)$.

We reject the null hypothesis at the 5% level of significance.

The p-value is 0.6315%. Have we encountered this p-value before? Yes, in the test of the hypothesis $\beta_1 = 0$ earlier using the test statistic $T_1 \sim t_8$, we arrived at the same p-value. Also for the test statistic $T_1 = 3.6694$. Note that $T_1^2 = (3.6694)^2 = 13.4646 = F$ above. Is this a coincidence? Not at all. In general, if $T \sim t_k$ has the t-distribution with k degrees of freedom, then $T^2 \sim F_{1,k}$. In this case, the ANOVA approach and the regression analysis give the same conclusion.

6.1.5 Regression Analysis in Excel

Being able to read the output from Excel enhances our understanding of the analysis and it serves as a check for the computations from other means or sources. We seek to define the output from Excel that gives the least-squares estimates of the parameters, test of significance of the parameters, their confidence intervals, ANOVA output as well as coefficient of determination.

The Excel spreadsheet has a regression function that provides most of the information and conclusion we discussed so far. Let us see the output for the income-education data in Figure 6.11.

Though there are convenient software packages for regression analysis, it is important for us to understand what the numbers mean and whether they make sense given the data set.

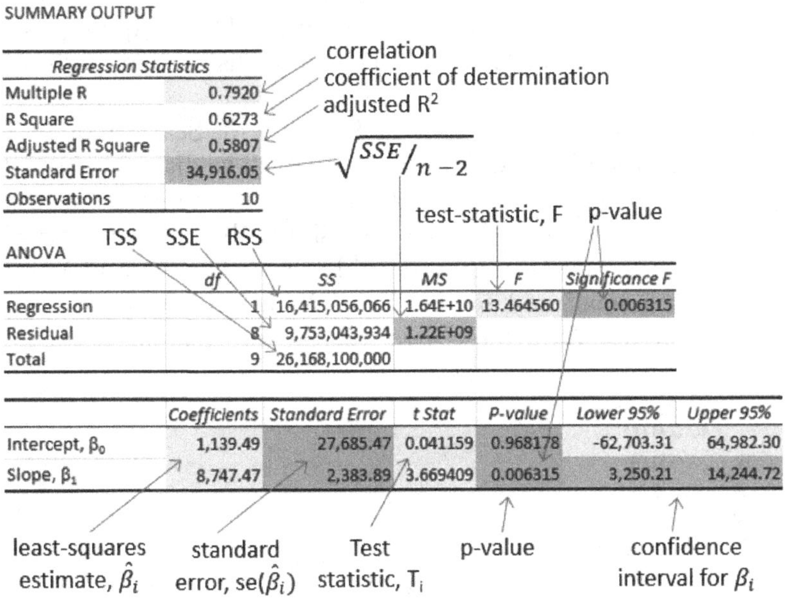

Figure 6.11: Sample excel output of income-education dataset.

6.1.6 *Checking the Residuals*

Examining the residuals or error terms from a regression model is an essential step. The residuals may reveal further information or structure that the linear model failed to capture. The outcome of the analysis may call into question the validity of the assumptions of the model. Further, the inspection and analysis of the residuals may provide valuable insight into the linear model and its inadequacies.

After the estimation of the parameters β_0 and β_1, in the regression model $Y = \beta_0 + \beta_1 X + \varepsilon$ we should compute the residuals $\varepsilon_i = Y_i - \hat{Y}_i$ and examine them to check that the assumptions about the error term ε holds. Do the residuals look like random white noise? For a given value of X, do the residuals or errors sum to 0 on average? Does the variance of ε look roughly the same for different values of X? When the errors have the same variance regardless of the independent variable X, we say the errors are homoscedastic (same variance). By looking or analyzing the residuals, we may find structures or further information in them to be extracted. For the income-education data, the following plot in Figure 6.12 gives the residuals. This is quite a small sample and the plot does not show any discernible pattern or feature.

Let us consider the data set in Figure 6.13; Scatterplot 2. The sample correlation between Y and X is 0.61. The diagram shows the least-squares regression line. Is a linear model suitable in this case?

It is a good practice to study the residuals by fitting a regression line or curve to the data. Figure 6.14 below shows the plot of the residuals. For small and big values of X, the residuals tend to be positive while for X in the central region, the residuals seem to be

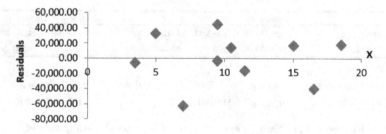

Figure 6.12: Income-education data residual plot.

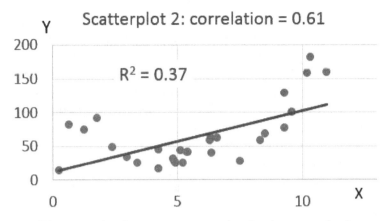

Figure 6.13: Least-square regression line in scatterplot 2.

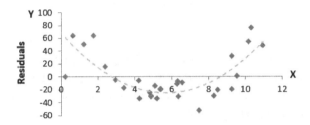

Figure 6.14: Scatterplot 2 residual plot.

negative. The residuals do not look like random noise. There may be a parabolic structure or information that can be extracted. So we need to revise or change the linear model.

6.1.7 *Difficulties with Linear Regression Models*

We highlight some of the common difficulties with the use of linear models and what may be done to mitigate them. To gain an understanding of how assumptions about the linear model may not hold in practice and how the choice of variables can affect the analysis.

1. Heteroscedasticity

This is a situation where the variances of the residuals or errors at different values of X are not the same. In the scatterplot below, we see that the variability of the points around the regression line seems

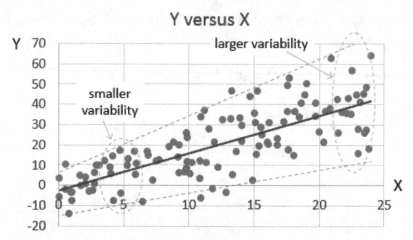

Figure 6.15: Graphical illustration of heteroscedasticity in the linear regression model.

to increase along with X. So the assumption of having a constant variance of the errors for different values X does not hold.

Figure 6.15 depicts a graphical illustration of heteroscedasticity in the linear regression model.

The issue of heteroscedasticity does not affect the computation of the least-squares estimators of the parameters. However, it does pose a problem for the estimation of the standard errors. To tackle this difficulty, we may use weighted least-squares or use robust standard errors.

Weighted least-squares: in the computation of the least-squares errors, points will be assigned different weights depending on where they come from and their variances. In this approach, we attempt to model the variance which in itself is a difficult task.

Robust standard errors: we use the residuals to obtain an estimate of the variance of the error term for various values of X.

2. Multicollinearity
This arises when some of the independent variables are closely correlated. It means that the least-squares estimates of the parameters will be unstable in that small perturbations in the observed values of the independent and dependent variables will result in large changes in the least-squares estimates. To overcome the problem, we should

remove the independent variables that are highly correlated with the others.

3. Stability of model parameters

Linear relationships such as those in the physical sciences tend to be stable over time. For example, if one measures the speed V at which a ball that is released from a height H hits the ground, we will find a linear relationship between $\ln V$ and $\ln H$, namely $\ln V = \frac{1}{2}\ln 2g + \frac{1}{2}\ln H$ and the coefficients in this equation are the same everywhere on earth and over time. The constant $g = 9.8\,\text{m/s}^2$ is known as the acceleration due to gravity. Linear relationships in finance, economics, and social sciences that exist may change over time for a variety of reasons. So we have to exercise great care in using linear models and to use new observations to test the validity of the linear model and estimated parameters.

Bibliography

Illowsky, B. and Dean, S. (2013). *Introductory Statistics*. OpenStax. pp. 679–763. Retrieved from openstar.org.

Triola, M. (2018). *Elementary Statistics*. 13th ed., Chaps. 10 and 12. Pearson, New Jersey, NJ.

6.2 Sample Questions

Please select the most appropriate response.

The following information pertains to Question 1 to Question 6.

An equity analyst studied the returns of UOB bank shares and the STI index over two weeks (ten trading days) in June 2020. A simple regression of UOB's return (in %) on STI's return (in %) gave the following result.

Regression Statistics	
Multiple R	0.9019
R Square	0.8134
Adjusted R Square	0.7901
Standard Error	0.0112
Observations	10

	Coefficients	Standard Error	t Stat	P-value (two-tailed test)
Intercept	0.0057	0.0038	1.5069	0.17027
STI (slope)	1.1478	0.1943	5.9061	0.00036

Question 1

What is the sample correlation between UOB's and STI's returns?

(a) 0.9019
(b) 0.8134
(c) 0.7901

Question 2

If one assumes that the linear relationship is stable, for a 1% increase in STI's return, what is the expected increase in UOB's return on average?

(a) 0.57%
(b) 1.1478%
(c) 0.9019%

Question 3

The analyst carried out a test of hypothesis to determine whether the slope of the regression line is significant using the test statistic, $t = 5.9061$. What is the distribution for this test statistic?

(a) The t-distribution with 9 degrees of freedom.
(b) The t-distribution with 8 degrees of freedom.
(c) The t-distribution with 1 degree of freedom.

Question 4

The analyst carried out a two-tail test of the null hypothesis that the slope of the regression line is equal to zero at the 1% level of significance. What was the outcome of the test?

(a) The test could not be carried out since the sample size of ten observations was too small.

(b) The test could be carried out but there is insufficient information given for us to conclude.
(c) The null hypothesis is rejected.

Question 5

To use the data to support certain claims about the Capital Asset Pricing Model (CAPM), the analyst would like to reject the null hypothesis that the intercept of the regression line β_0 is equal to zero. Which of the following criteria or practices would result in the desired outcome?

  (i) Set the level of significance at 10% for a two-tailed test.
 (ii) Set the level of significance at 20% for a two-tailed test.
(iii) Set the alternative hypothesis to be $\beta_0 > 0$ and the level of significance at 10%.
(iv) Set the alternative hypothesis to be $\beta_0 < 0$ and the level of significance at 10%.

(a) Only (i) and (ii)
(b) Only (ii) and (iii)
(c) Only (ii) and (iv)

Question 6

What is the coefficient of determination for the regression?

(a) 90.19%
(b) 81.34%
(c) 79.01%

The following information pertains to Questions 7 to 11.
    A researcher believes that the short-term interest rate is one of the factors that drive stock market returns and that the relationship is linear. She analyses the monthly returns of the stock market and the short-term interest rates over five years (60 observations). The result of the simple regression of the stock market returns (in %) on the short-term interest rate (in %) is given in the following table.

ANOVA					
	df	*SS*	*MS*	*F*	*Significance F*
Regression	1	0.0972	0.0972	6.2627	0.0152
Residual	58	0.8999	0.0155		
Total	59	0.9971			

	Coefficient	*Standard Error*	*t Stat*	*P-value (two-tailed test)*	*Lower 95%*	*Upper 95%*
Intercept	11.98%	0.0296	4.0458	0.0002	0.0605	0.1791
Interest rate (slope)	−2.5529	1.0201	−2.5025	0.0152	−4.5949	−0.5109

Question 7

What is the coefficient of determination?

(a) Regression sum of squares = 0.0972
(b) $0.0155/0.0972 = 0.1597$
(c) $0.0972/0.9971 = 0.0975$

Question 8

What is the sample correlation?

(a) $\sqrt{0.0972} = 0.3117$.
(b) $\sqrt{0.0972/0.9971} = 0.3122$.
(c) None of the above.

Question 9

The researcher performed an F-test using the output in the table. What are the conclusions of the F-test?

(i) The intercept of the regression line is significantly different from zero at the 5% level of significance.
(ii) Both the intercept and the slope of the regression line are significantly different from zero at the 5% level of significance.
(iii) The slope of the regression line is significantly different from zero at the 5% level of significance but not at the 1% level of significance.

(a) The F-test allows us to conclude (i) only.
(b) The F-test allows us to conclude (ii) only.
(c) The F-test allows us to conclude (iii) only.

Question 10

Assuming the results of the regression analysis are valid, what will be the effect on the equity market when the short-term interest rate goes up by 1%?

(a) There is no significant relationship between the short-term interest rate and the return on the equity market.
(b) Return on equity market will be higher by 11.98% on average.
(c) Return on equity market will be lower by 2.55% on average.

Question 11

The sales manager of an equity fund wants to use the results of the regression analysis for marketing equity funds. In particular, she would like to claim that each percentage point drop in the short-term interest rate will result in equity returns being higher by 4.5% on average. Given that the US central bank (the Federal Reserve) is expected to cut the Federal Funds rates several times in the next twelve months, it is sensible to invest in equities if the claim is valid. She carries out a two-tailed test of the null hypothesis, H_0: $\beta_1 = -4.5$ versus the alternative hypothesis, H_A: $\beta_1 \neq -4.5$ at the 5% level of significance. What is the conclusion of the test?

(a) There is insufficient information provided to arrive at a conclusion.
(b) The null hypothesis is rejected.
(c) The null hypothesis is not rejected.

Solutions

Question 1

Solution: Option **a** is correct.

The multiple R gives the correlation between the two variables in a simple linear regression.

Question 2

Solution: Option **b** is correct.

This is given by the coefficient associated with the STI index, which is 1.1478.

Question 3

Solution: Choice **b** is correct.

The test statistic has a t-distribution with $10 - 2 = 8$ degrees of freedom.

Question 4

Solution: Option **c** is correct.

The p-value of the test is 0.36% which is much smaller than 1%. Hence the rejection.

Question 5

Solution: Option **b** is correct.

The test statistic $t = 1.5069$ has a p-value of 17.027%. Hence a 20% level of significance will lead to the rejection of the null hypothesis. From the p-value of 17.027% (two-tailed), we see that the probability that $t > 1.5069$ is $17.027\%/2 = 8.514\%$. Hence the setting in (iii) will lead to the rejection of the null hypothesis.

Question 6

Answer: Option **b** is correct.

According to the table, the R square is 0.8134.

Question 7

Solution: Option **c** is correct

By definition, $R^2 = 0.0972/0.9971 = 0.0975$.

Question 8

Solution: Option **c** is correct

In general, the sample correlation is the square root of R^2. But we must be careful about the sign. Since the estimate for the coefficient is negative, the sample correlation is -0.3122.

Question 9

Solution: Option **c** is correct

In the simple linear regression, the F-test is equivalent to the t-test for the coefficient of the independent variable. Since the p-value is 1.52%, we have a conclusion of iii.

Question 10

Solution: Option **c** is correct.

Since the coefficient is -2.5529, the return will be lower by 2.55% on average.

Question 11

Solution: Option **c** is correct

Since the 95% CI for β_1 is $(-4.5949, -0.5109)$ and it contains the value -4.5, the null hypothesis will not be rejected at the 5% level of significance.

PART III
Quantitative Methods

PART II
Quantitative Method

Chapter 7

Boolean Algebra and Logic Gates

Boolean algebra is the mathematical foundation of digital circuits. Boolean algebra specifies the relationship between Boolean variables that are used to design combinational logic circuits using logic gates.

7.1 Learning Objectives

- Define Boolean algebra and logic gates and the conceptual framework.

7.2 Main Takeaways

7.2.1 *Main Points*

- Boolean algebra is the branch of algebra in which the values of the variables are the truth values true and false, usually denoted 1 and 0 respectively.
- The primary modern use of Boolean algebra is in computer programming languages.
- Logic gates are electronic circuits that implement the basic operations of Boolean algebra. There is a symbol for each gate.

7.2.2 *Main Terms*

- **Algorithm:** A process or set of rules to be followed in calculations or other problem-solving operations, especially by a computer.

- **Associative Law:** Either of two laws relating to operations of addition and multiplication, stated symbolically: $a + (b + c) = (a + b) + c$, and $a(bc) = (ab)c$, that is, the terms or factors may be associated in any way desired.
- **Commutative Law:** Either of two laws relating to operations of addition and multiplication, stated symbolically: $a + b = b + a$ and $ab = ba$. From these laws, it follows that any finite sum or product is unaltered by reordering its terms or factors.
- **Binary Variables:** Binary variables are *variables* that only take two values. For example, male or female, true or false, and yes or no.

7.3 Boolean Algebra

Boolean algebra provides a framework for studying the logic of statements or variables and operations on them. It has applications in the design of electronic circuits used in computational devices.

Every statement or variable has two possible values that correspond to "true" and "false" and this mimics an electrical circuit where the current could be "on" or "off". All operations on statements or variables may be derived from just three operations AND, OR, and NOT.

Boolean algebra is named after the English mathematician George Boole. Variables are usually denoted by letters of the alphabet; $A, B, C, \ldots, Z$. Each variable takes on two possible values: true (=1) or false (=0). As such, these variables are termed binary variables.

We may combine these variables using operations such as:

- AND operation denoted by $\wedge$ or . and it may be regarded as "multiplication"
- OR operation denoted by $\vee$ or $+$ and it may be regarded as "addition"
- NOT operation denoted using the prime symbol or an overhead bar A and it may be regarded as negation.

A.B		A	B	A ∧ B
1.1 = 1	⟶	1	1	1
1.0 = 0	⟶	1	0	0
0.1 – 0	⟶	0	1	0
0.0 = 0	⟶	0	0	0

Figure 7.1: Truth table for AND operation.

7.3.1 Boolean Operations and Laws

7.3.1.1 AND Operation

7.3.1.1.1 Definition

If A and B are variables, then $A \wedge B$ is again a variable. The truth value of $A \wedge B$ depends on the truth value of A and of B and is defined as follows:

- $A \wedge B$ is true (value $= 1$) only when both A and B are true.
- $A \wedge B$ is false (value $= 0$) as long as A or B or both are false.

Using the Truth table in Figure 7.1, we can think of $A \wedge B$ as the multiplication $A.B$ as demonstrated in the figure.

7.3.1.1.2 AND Operation: The Intuition

We may think of variables as statements. For example:

A: Alice is a student of University U.
B: Bob likes bubble tea.

A could be true (1) or false (0) and this applies to B too. Now $A \wedge B$: Alice is a student of University U and Bob likes bubble tea. The statement, $A \wedge B$ is true when both statements "Alice is a student of University U" and "Bob likes bubble tea" are true. As long as one of A or B or both are false then the statement $A \wedge B$ is false.

Easy to check that the AND operation is commutative: $A \wedge B = B \wedge A$.

A+B		A	B	A ∨ B
1 + 1 = 1	⟶	1	1	1
1 + 0 = 1	⟶	1	0	1
0 + 1 = 1	⟶	0	1	1
0 + 0 = 0	⟶	0	0	0

Figure 7.2: Truth table for OR operation.

7.3.1.2 *OR Operation*

7.3.1.2.1 Definition

Given variables A and B, $A \vee B$ is again a variable. The truth value of $A \vee B$ is defined below:

- $A \vee B$ is true (value = 1) as long as one of A or B or both are true.
- $A \vee B$ is false (value = 0) if both A and B are false.

Using the Truth table in Figure 7.2, we can think of $A \vee B$ as the addition $A.B$ as demonstrated in the figure.

Note: 1 is the largest possible value for $A + B$.

7.3.1.2.2 OR Operation: The Intuition

Consider the earlier statements.

A: Alice is a student of University U.
B: Bob likes bubble tea.

$A \vee B$: Alice is a student of University U or Bob likes bubble teas. The statement, $A \vee B$ is true whenever one or both of the statements "Alice is a student of University U" and "Bob likes bubble tea" are true. The version of "or" in daily usage is the exclusive version: "would you like coffee or tea" means you choose one and not both. In Boolean algebra, the "or" operation is inclusive: as long as one of the variables is true or both are true, $A \vee B$ is true.

Note that $\vee$ is commutative: $A \vee B = B \vee A$.

7.3.1.3 *NOT Operation*

7.3.1.3.1 Definition

The NOT operation or negation applies to one variable. If A is a variable, the negation of A, denoted as A' or $\overline{A}$, is true whenever A is false and conversely. This is shown in the Truth table in Figure 7.3.

A	A'
1	0
0	1

Figure 7.3: Truth table for NOT operation.

Suppose A is the following statement.

A: "Alice is a student of University U".

Then A' the negation of A is the statement "Alice is not a student of University U" or "It is not the case that Alice is a student of University U".

7.3.1.4 *Equivalent Statements or Variables*

Suppose $f(A, B, C)$ and $g(A, B, C)$ are variables that depend on variables A, B, and C. How can we check that they are equivalent?

As long as they have the same truth values under all assignments of truth values for A, B, and C, we regard $f(A, B, C)$ and $g(A, B, C)$ as equivalent.

7.3.1.5 *Examples*

(a) Let $f(A, B, C) = (A \vee B) \wedge C$ and $g(A, B, C) = (A \wedge C) \vee (B \wedge C)$. Is $(A \vee B) \wedge C = (A \wedge C) \vee (B \wedge C)$?

Solution:

A	B	C	A ∨ B	(A ∨ B) ∧ C	A ∧ C	B ∧ C	(A ∧ C) ∨ (B ∧ C)
1	1	1	1	1	1	1	1
1	1	0	1	0	0	0	0
1	0	1	1	1	1	0	1
1	0	0	1	0	0	0	0
0	1	1	1	1	0	1	1
0	1	0	1	0	0	0	0
0	0	1	0	0	0	0	0
0	0	0	0	0	0	0	0

We check that the columns for $(A \vee B) \wedge C$ and $(A \wedge C) \vee (B \wedge C)$ have the same truth values under all assignments of truth values for A, B, and C. Thus they are equivalent.

Recall the distributive law for numbers: $(a + b) \times c = (a \times c) + (b \times c)$. It also holds for $(A \vee B) \wedge C = (A \wedge C) \vee (B \wedge C)$ where $\vee$ is regarded as addition $(+)$ and $\wedge$ is taken as multiplication $(\times)$.

(b) Is $(A \wedge B)' = A\prime \vee B'$?

Solution: There are two possible assignments of the truth value for A as well as for B. So there are altogether $2 \times 2 = 4$ possible assignments that we have to check in order to verify the validity or otherwise of the equation. In general, if a statement or function contains n distinct variables, there are 2^n assignments to check.

A	B	$A \wedge B$	$(A \wedge B)'$	$A' \vee B'$
1	1	1	0	0
1	0	0	1	1
0	1	0	1	1
0	0	0	1	1

From the above table, we see that $(A \wedge B)' = A' \vee B'$. It can also be shown that $(A \vee B)' = A' \wedge B'$. These relations are known as De Morgan's laws.

(c) Is $(A \vee B) \wedge C = A \vee (B \wedge C)$?

Solution:

A	B	C	$A \vee B$	$(A \vee B) \wedge C$	$B \wedge C$	$A \vee (B \wedge C)$
1	1	1	1	1	1	1
1	1	0	1	0	0	1
1	0	1	1	1	0	1
1	0	0	1	0	0	1
0	1	1	1	1	1	1
0	1	0	1	0	0	0
0	0	1	0	0	0	0
0	0	0	0	0	0	0

There are two instances where $(A \vee B) \wedge C$ and $A \vee (B \wedge C)$ don't have the same truth values, namely when $A = B = 1$, $C = 0$ and when $A = 1$, $B = C = 0$. Hence, they are not equivalent. So $(A \vee B) \wedge C \neq A \vee (B \wedge C)$.

A	B	A ⇒ B
1	1	1
1	0	0
0	1	1
0	0	1

A	B	A ⇐ B
1	1	1
1	0	1
0	1	0
0	0	1

Figure 7.4: Truth table for imply operations.

Imply operation, $A \Rightarrow B$: if A then B or A implies B

The truth table for the imply operation shown in Figure 7.4, usually denoted as $\Rightarrow$, is as follows in the figure.

Note: $A \Rightarrow B \neq A \Leftarrow B$ but $A \Rightarrow B = B \Leftarrow A$. So the operation $\Rightarrow$ is not commutative.

Consider the earlier statements:

A: Alice is a student of University U.
B: Bob likes bubble tea.

$A \Rightarrow B$: if Alice is a student of University U then B likes bubble tea. Notice that from the truth table, the only time that $A \Rightarrow B$ is false (=0) is when A is true (=1) and B is false. That is if Alice is a student of University U, but Bob does not like bubble tea, then the statement $A \Rightarrow B$ is false.

Here A is the premise. If we start with a false premise, that is A is false then the statement $A \Rightarrow B$ is automatically true. A false premise can imply any B; and the whole statement $A \Rightarrow B$ is considered true.

7.3.1.6 *Two Special Variables 0 and 1*

0 is the variable that always takes the value 0 (always false). 1 is the variable that always takes the value 1 (always true). This is shown in the Truth table for special variables in Figure 7.5.

Note: $0' = 1$. Also A $\vee$ 0 = A. A $\wedge$ 0 = 0. And A $\vee$ 1 = 1; A $\wedge$ 1 = A

Properties

For any variables A, B, and C, we have:
Associative Law

$$A \wedge (B \wedge C) = (A \wedge B) \wedge C$$
$$A \vee (B \vee C) = (A \vee B) \vee C$$

A	0	A∨0		A	0	A∧0		A	1	A∨1		A	1	A∧1
1	0	1		1	0	0		1	1	1		1	1	1
0	0	0		0	0	0		0	1	1		0	1	0

Figure 7.5: Truth table for special variables.

Commutative Law

$$A \vee B = B \vee A$$

$$A \wedge B = B \wedge A$$

Distributive Law

$$A \vee (B \wedge C) = (A \vee B) \wedge (A \vee C)$$

$$A \wedge (B \vee C) = (A \wedge B) \vee (A \wedge C)$$

Absorption: $A \wedge (A \vee B) = A$ and $A \vee (A \wedge B) = A$

$A \vee 0 = A$, $A \wedge 1 = A$, $A \vee A = A$, $A \wedge A = A$, $(A')' = A$

$A \vee 1 = 1$, $A \wedge 0 = 0$, $A \vee A' = 1$, $A \wedge A' = 0$

7.4 Logic Gates

The results of the operations in Boolean algebra may be implemented in electronic circuits using devices known as logic gates. The variables having only two truth values 0 or 1 is like the "on" or "off" condition in a circuit. See the introductory diagram to logic gates in Figure 7.6.

Buffer: output $P = A$

7.4.1 *Types of Logic Gates*

7.4.1.1 *NOT, AND, OR, NAND, NOR*

The circuit symbols for the operations are shown in Figure 7.7.

Note: *NAND* comes from "Not AND" and *NOR* is from "Not OR".

7.4.1.2 *EXOR or exclusive OR*

EXOR: output $P = A \oplus B$ is defined below and its Truth table is shown in Figure 7.8. The variable $A \oplus B = 1$ if and only if $A \neq B$.

Note: $A \oplus B = (A \vee B) \wedge (A' \vee B')$.

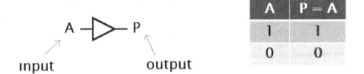

A	P = A
1	1
0	0

input   output

Figure 7.6: Introductory diagram to logic gates.

Not: output P = A' or $\bar{A}$ A —▷o— P

A	P = A'
1	0
0	1

AND: output P = A ∧ B $\frac{A}{B}$ ⊐D— P

OR: output P = A ∨ B $\frac{A}{B}$ ⊐D— P

NAND: output P = (A ∧ B)' = A' ∨ B' $\frac{A}{B}$ ⊐D— P

NOR: output P = (A ∨ B)' = A' ∧ B' $\frac{A}{B}$ ⊐D— P

Figure 7.7: Circuit symbols for different types of logic gates.

A	B	P = A ⊕ B
1	1	0
1	0	1
0	1	1
0	0	0

Figure 7.8: Truth table for EXOR operations.

7.4.1.3 *EXNOR or Exclusive NOR*

EXNOR: output $P = (A \oplus B)'$ or $\overline{A \oplus B}$ is defined below. The variable $(A \oplus B)' = \overline{A \oplus B} = 1$ if and only if $A = B$. The Truth table for EXNOR is shown in Figure 7.9.

Note: $(A \oplus B)' = (A \wedge B) \vee (A' \wedge B')$.

Summary: Truth table for the operations (Figure 7.10)

Note that all other operations can be expressed using AND, OR, and NOT.

A	B	P = (A ⊕ B)'
1	1	1
1	0	0
0	1	0
0	0	1

Figure 7.9: Truth table for EXNOR operations.

A	B	A ∧ B AND	A ∨ B OR	A' NOT	(A ∧ B)' NAND	(A ∨ B)' NOR	A ⊕ B XOR	(A ⊕ B)' EXNOR
1	1	1	1	0	0	0	0	1
1	0	0	1	0	1	0	1	0
0	1	0	1	1	1	0	1	0
0	0	0	0	1	1	1	0	1

Figure 7.10: Summary of truth table for all the operations.

Bibliography

Mehta, V. (2015). *Logic Gates for Beginners.* GRIN Publishing, Munich, Germany.

7.5 Sample Questions

Please select the most appropriate response.

Question 1

In Boolean algebra, which of the following properties does the OR operation ∨ (+) have?

 (i) Associative property: $A \vee (B \vee C) = (A \vee B) \vee C$
 (ii) Commutative property: $A \vee B = B \vee A$
(iii) Distributive property: $A \vee (B \wedge C) = (A \vee B) \wedge (A \vee C)$

(a) Associative and commutative properties
(b) Associative and distributive properties
(c) Associative, commutative, and distributive properties

Question 2

Which of the following expresses the absorption law?

(a) $A + AB = A$
(b) $A + AB = B$
(c) $AB + AA' = A$

Question 3

The Boolean function $A+BC$ is a reduced form of _____

(a) $AB + BC$
(b) $(A + B)(A + C)$
(c) $A'B + AB'C$

Question 4

The output of a logic gate is 1 when both the inputs have truth value 0 as shown below:

Input		Output	Input		Output
A	B	C	A	B	C
0	0	1	0	0	1
0	1	0	0	1	0
1	0	0	1	0	0
1	1	0	1	1	1

The gates are _____

(a) a NAND gate and an EX-OR gate respectively
(b) an OR gate and an EX-NOR gate respectively
(c) a NOR gate and an EX-NOR gate respectively

Question 5

Which of the following are known as universal gates in that any digital circuit can be made using these gates? That is, they can replicate the functions of all the other gates.

(a) NAND and NOR
(b) AND and OR
(c) XOR and OR

Question 6

Consider the circuit below.

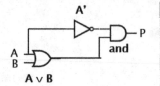

A	B	P
1	1	X
1	0	
0	1	Y
0	0	

What are X and Y of the output P?

(a) $X = 1, Y = 1$
(b) $X = 0, Y = 1$
(c) $X = 0, Y = 0$

Solutions

Question 1

Solution: Option **c** is correct

Question 2

Solution: Option **a** is correct

Explanation: The expression for Absorption Law is given by A + AB = A.

Proof: $A + AB = A(1+B) = A$ (since $1+B = 1$ as per 1's Property).

Question 3

Solution: Option **b** is correct

Explanation:

$$
\begin{aligned}
(A + B)(A + C) &= AA + AC + BA + BC \\
&= A + AC + AB + BC \, (\text{by commutative property}) \\
&= A(1 + C + B) + BC \\
&= A + BC
\end{aligned}
$$

$(1 + B + C = 1 \text{ by } 1's\, Property)$.

Question 4

Solution: Choice **c** is correct

Explanation: The output of a logic gate is 1 when all inputs are at logic 0. The gate is NOR. The output of a logic gate is 1 when all inputs are at logic 0 or all inputs are at logic 1, then it is EX-NOR. (The truth tables for NOR and EX-NOR Gates are shown in the above table).

Question 5

Solution: Choice **a** is correct

Explanation: The NAND and NOR gates are known as universal gates because any digital circuit can be realized completely by using these two gates, and also they can generate the three basic gates

AND, OR, and NOT. For example, to replicate NOT A, we see that A NOR $A = (A \vee A)' = A'$. To replicate AND, we have

$$[(A \text{ NAND } B) \text{ NOR } (A \text{ NAND } B)] = (A \wedge B)' \text{ NOR } (A \wedge B)' = A \wedge B.$$

Also, $[(A \text{ NOR } B) \text{ NOR } (A \text{ NOR } B)] = (A \text{ NOR } B)' = A \vee B.$

Question 6

Solution: Choice **b** is correct

(b) $X = 0, Y = 1$

$$P = A' \wedge (A \vee B)$$
$$= (A' \wedge A) \vee (A' \wedge B)$$
$$= 0 \vee (A' \wedge B) = A' \wedge B$$

Chapter 8

Number System

8.1 Number System: Bases

A *number base* is the number of digits or a combination of digits that a system of counting uses to represent numbers. A base can be any whole number greater than 1.

8.1.1 *Learning Objectives*

- Convert any number given in base A to one in base B.

8.1.2 *Main Takeaways*

8.1.2.1 *Main Points*

- The most commonly used number system is the decimal system, commonly known as the base 10 system.
- The base of any number may be written beside the number in the subscript. For example, 17_8 is read as 17 base 8, which is equal to 15_{10} in base 10.

8.1.2.2 *Main Terms*

- **Octal System:** The octal numeral system, or oct for short, is the base-8 number system and uses the digits 0 to 7.

- **Hexadecimal System:** In mathematics and computing, hexadecimal is a positional system that represents numbers using a base of 16. Unlike the common way of representing numbers with ten symbols, it uses sixteen distinct symbols, most often the symbols "0"–"9" are used to represent values zero to nine, and "A"–"F" is used to represent values 10 to 15 respectively.

8.1.3 *Different Bases*

The number system that we use for daily transactions or activities is the base 10 system. For mechanical devices, a number system in other bases such as base 2 or base 16 may be more useful in the sense that computations are made easier (base 2) or that less storage space is needed to represent a number (base 16).

For a given positive integer, $A > 1$, every number N can be expressed uniquely as a sum of powers of A. The coefficients of the terms in the expression then give a representation of the number N in the base A system. One can convert any number given in base A to one in base B by first expressing the number in the common base 10 system.

8.1.3.1 *Base 10*

We use the base 10 number system in daily activities. In a base 10 system, we need 10 symbols to denote quantities, namely, 0, 1, 2, 3, 4, 5, 6, 7, 8, 9.

We write:

$$139 = 100 + 30 + 9 = 1 \times 10^2 + 3 \times 10^1 + 9 \times 10^0$$

We are familiar with $4 + 3 = 7$ and $4 + 9 = 13$. When the sum $4 + 9 = 13 = 10 + 3 \geq 10$, is greater or equal to 10, there is a "carry over" of 1 to one order higher.

8.1.3.2 *Base 2*

In the base 2 system, we need two symbols 0 and 1. To indicate the base 2 system, we add a subscript 2 to the number. We compare numbers in base 10 and base 2.

$$0_{10} = 0_2$$

$$1_{10} = 1_2$$

$$2_{10} = 1 \times 2^1 + 0 \times 2^0 = 10_2$$

$$3_{10} = 2_{10} + 1_{10} = 1 \times 2^1 + 1 \times 2^0 = 11_2$$

$$4_{10} = 1 \times 2^2 + 0 \times 2^1 + 0 \times 2^0 = 100_2$$

$$6_{10} = 4_{10} + 2_{10} = 1 \times 2^2 + 1 \times 2^1 + 0 \times 2^0 = 110_2$$

$$27_{10} = 16_{10} + 8_{10} + 2_{10} + 1_{10}$$

$$= 1 \times 2^4 + 1 \times 2^3 + 0 \times 2^2 + 1 \times 2^1 + 1 \times 2^0 = 11011_2$$

To express a number in base 10 as a number in base 2, we first write it as a sum of powers of 2 where the coefficients are 0 or 1. This expression is unique.

Example

$$103_{10} = 64_{10} + 32_{10} + 4_{10} + 2_{10} + 1_{10}$$

$$= 2^6 + 2^5 + 2^2 + 2^1 + 2^0 = 1100111_2$$

To convert from base 2 to base 10, we reverse the process. For example,

$$100100101_2 = 2^8 + 2^5 + 2^2 + 2^0 = 256 + 32 + 4 + 1 = 293_{10}$$

With addition in base 2, if $a_2 + b_2 \geq 2$, then there is a "carry over" of 1 to one order higher.

Example

$$
\begin{array}{r}
1100111_2 \\
+ \quad 110110_2 \\
\hline
10011101_2
\end{array}
$$

Exercise

What is 10011101_2 in base 10?

(a) 103_{10}
(b) 141_{10}
(c) 157_{10}

Solution: Choice **c** is correct.

$$10011101_2 = 2^7 + 2^4 + 2^3 + 2^2 + 2^0$$
$$= 128_{10} + 16_{10} + 8_{10} + 4_{10} + 1_{10} = 157_{10}$$

8.1.3.3   *Base 8: Octal System*

In the base 8 system, we need eight symbols namely 0, 1, 2, 3, 4, 5, 6, and 7. Again to write a number in base 10 to one in base 8, we first express it uniquely as a sum of the powers of 8 where the coefficients range from 0 to 7.

Example

$$689_{10} = 1 \times 8^3 + 2 \times 8^2 + 6 \times 8^1 + 1 \times 8^0 = 1{,}261_8$$

We can go from the base 8 (octal system) to the base 10 system by summing the product of the digits and their corresponding powers of 8.

Example

$$214003_8 = 2 \times 8^5 + 1 \times 8^4 + 4 \times 8^3 + 3 \times 8^0 = 71{,}683_{10}$$

8.1.3.4   *Base 16: Hexadecimal System*

The base 16 number system requires 16 symbols that are usually denoted as 0, 1, 2, 3, 4, 5, 6, 7, 8, 9, A, B, C, D, E, and F. Here $A \equiv 10, B \equiv 11, C \equiv 12, D \equiv 13, E \equiv 14, F \equiv 15$. The computation works the same way as the other bases.

Example

As an example, let us convert a base 16 number to one in base 10:

$$2\text{EB7}_{16} = 2 \times 16^3 + E \times 16^2 + B \times 16^1 + 7 \times 16^0$$
$$= 2 \times 16^3 + 14 \times 16^2 + 11 \times 16^1 + 7 \times 16^0$$
$$= 11{,}959_{10}$$

Exercise

Suppose $1011101_2 = 1T5_8$
What is the value of T?

(a) $T = 0$
(b) $T = 3$
(c) $T = 5$

Solution: Choice **b** is correct.

$$1011101_2 = 64 + 16 + 8 + 4 + 1 = 93$$
$$1T5_8 = 1 \times 8^2 + T \times 8^1 + 5 = 69 + T \times 8 = 93$$
$$T = 3$$

Bibliography

Rosen, K. (2010). *Elementary Number Theory And Its Applications*. 6th ed., Chaps. 1 and 2. Pearson, New Jersey, NJ.

8.2 Sample Questions

Please select the most appropriate response.

Question 1
Convert 235_8 into base 10.

(a) 137_{10}
(b) 157_{10}
(c) 210_{10}

Foundations for Fintech

Question 2

Which of the following digit does is not appear in the Octal representation of a number?

(a) 0
(b) 4
(c) 8

Question 3

Which of the following numbers is equal to $2{,}345_{10}$?

(a) $92B_{16}$
(b) 4453_8
(c) 100100101001_2

Question 4

Let $V = 111001100_2$, $W = 1230_7$, $X = 714_8$, $Y = 1CE_{16}$, $Z = 462_{10}$.
Consider the following statements:

(i) $V = W$
(ii) $W = X$
(iii) $X < Y$
(iv) $V + Z = W + X$

Which of the above statements from (i) to (iv) are correct?

(a) Only (i) and (iv) are correct
(b) Only (iii) is correct
(c) None of the above

Solutions

Question 1

Solution: Choice **b** is correct

Explanation: Remember: any nonzero number to the zero power equals one. In this problem, you have

$$5 \times 8^0 = 5$$
$$3 \times 8^1 = 24$$
$$2 \times 8^2 = 128$$

Now simply add these together, $5 + 24 + 128 = 157_{10}$

Question 2

Solution: Option **c** is correct

Explanation: Octal numbers are formed with a combination of digits from 0 to 7 only.

Question 3

Solution: Option **c** is correct

Explanation:

$$92B_{16} = 9 \times 16^2 + 2 \times 16^1 + B \times 16^0$$
$$= 2{,}304 + 32 + 11 = 2{,}347$$
$$4453_8 = 4 \times 8^3 + 4 \times 8^2 + 5 \times 8^1 + 3 \times 8^0$$
$$= 2{,}048 + 256 + 40 + 3 = 2{,}347$$
$$100100101001_2 = 1 \times 2^{11} + 1 \times 2^8 + 1 \times 2^5 + 1 \times 2^3 + 1 \times 2^0$$
$$= 2{,}048 + 256 + 32 + 8 + 1 = 2{,}345$$

Question 4

Solution: Option **c** is correct

Explanation:

$$V = 111001100_2 = 1 \times 2^8 + 1 \times 2^7 + 1 \times 2^6$$
$$+ 0 \times 2^5 + 0 \times 2^4 + 1 \times 2^3 + 1 \times 2^2 + 0 \times 2^1 + 0 \times 2^0$$
$$= 256 + 128 + 64 + 8 + 4 = 460_{10}$$
$$W = 1230_7 = 1 \times 7^3 + 2 \times 7^2 + 3 \times 7^1 + 0 \times 7^0$$
$$= 343 + 98 + 21 + 0 = 462_{10}$$
$$X = 714_8 = 7 \times 8^2 + 1 \times 8^1 + 4 \times 8^0$$
$$= 448 + 8 + 4 = 460_{10}$$
$$Y = 1CE_{16} = 1 \times 16^2 + C \times 16^1 + E \times 16^0$$
$$= 1 \times 16^2 + 12 \times 16^1 + 14 \times 16^0 = 256 + 192 + 14 = 462_{10}$$

Thus $V = X = 460$ and $W = Y = Z = 462$ and only (iii) and (iv) are correct.

Chapter 9

Modular Arithmetic

9.1 Modular Arithmetic

Modular arithmetic is a system of arithmetic for integers, which considers the remainder of each number for a given fixed integer. In modular arithmetic, numbers "wrap around" upon reaching a given fixed quantity (this given quantity is known as the modulus) to leave a remainder.

9.1.1 *Learning Objectives*

- Computing addition and multiplication tables associated with modular arithmetic.

9.1.2 *Main Takeaways*

9.1.2.1 *Main Points*

- An intuitive example of modular arithmetic is a 12-hour clock. If it is 10:00 now, then after 5 hours the clock will show 3:00 instead of 15:00. 3 is the remainder when 15 is divided by 12. We say or write $15 = 3$ modulo 12.
- A prime number (or a prime) is a natural number greater than 1 that is not a product of two smaller natural numbers. A natural number greater than 1 that is not prime is called a composite number. For natural numbers m and n, we say that m divides n if

there exists a natural number k such that $n = km$. We say that m is a divisor or a factor of n. We write $m|n$ to indicate m divides n.

- The Euclidean algorithm is an efficient method for computing the greatest common divisor (GCD) of two integers (numbers). The GCD of two integers is the largest number that divides them both.
- The Euclidean algorithm is based on the principle that the GCD of two numbers does not change when the larger number is replaced by its difference with the smaller number.

9.1.2.2 *Main Terms*

- **Integer:** An integer (from the Latin integer meaning "whole") is colloquially defined as a number that can be written without a fractional component.
- **Additive Inverse:** In mathematics, the additive inverse of a number **a** is the number that, when added to **a**, yields zero. This number is also known as the opposite (number), sign change, and negation and is usually written as $-$ **a**. For a real number, it reverses its sign: the opposite to a positive number is negative, and the opposite to a negative number is positive. For example -3 is the additive inverse of 3 since $3 + (-3) = 0$.
- **Polynomials:** In mathematics, a polynomial is an expression consisting of variables (also called indeterminates) and coefficients, that involves only the operations of addition, subtraction, multiplication, and non-negative integer exponents of variables.

Finding the divisors of numbers and modular arithmetic have applications in computer science, cryptography, and mathematics. The Euclidean algorithm, which is more than 2,000 years old and its extended version are reliable recipes for determining the GCD between two numbers. This section serves as an introduction to these topics.

The GCD between two integers can be written as an integer combination of the two integers. Computing the addition and multiplication tables associated with modular arithmetic allows us to investigate its structure.

9.1.3 *Positive Integers and Prime Numbers*

The positive integers are 1, 2, 3, 4, 5, 6, ... Some integers may be expressed as products of other integers. For example, $72 = 2 \times 3 \times 12$. We say that 2, 3, 12 are divisors or factors of 72. As $72 = 1 \times 72$, so 1 and 72 are naturally factors of 72. Can you list all the factors of 72? Note that 72 has 12 factors (counting positive integers only).

A positive integer $P > 1$ is prime if its factors are 1 and P itself and no others. We work with factors that are positive integers only. For example, 2, 3, 5, 7, 11, 13, 17, 19, 23, 29, ..., are primes. Primes are the building blocks of the integers.

9.1.3.1 *Common Divisors*

Let d and n be positive integers. If $n = k \times d$ for some integer k, we say that n is a multiple of d (as well as k). The integers d and k are factors of n. We say that d divides n, and this is denoted as $d|n$. If $d|n$ and $d|m$, we say that d is a common factor of n and m. For example, three is a common factor of 30 and 72. 2 is also a common factor of 30 and 72. It is easy to check that the greatest common factor of $30 = 6 \times 5$ and $72 = 6 \times 12$ is 6.

Let m and n be positive integers, the GCD of m and n is an integer d such that $d|m$ and $d|n$, and for every common divisor t of m and n we have $t|d$. The GCD of m and n is denoted by $\gcd(m, n)$. Note that $\gcd(m, n) = \gcd(n, m)$ by definition.

Given m and n, how can we find $\gcd(m, n)$? This is conceptually easy: write each of m and n as a product of all its prime factors and we can read off the gcd. For example, to find $\gcd(107{,}907{,}800, 94{,}128{,}804)$ we write them as product of prime factors:

$$107{,}907{,}800 = 2^3 \times 3^0 \times 5^2 \times 7^3 \times 11^2 \times 13^1$$

$$94{,}128{,}804 = 2^2 \times 3^4 \times 5^0 \times 7^4 \times 11^2 \times 13^0$$

$$\text{So, gcd} = 2^2 \times 3^0 \times 5^0 \times 7^3 \times 11^2 \times 13^0 = 166{,}012$$

The preceding method though conceptually easy is quite tedious and time-consuming in practice. The factorization of big numbers was used in cryptography for this reason. Do we have a systematic way of finding $\gcd(m, n)$? The answer is the Euclidean algorithm.

9.1.4   *Euclidean Algorithm*

Let m and n be integers with $m \geq n$. Let $\gcd(m, n) = d$.

Write $m = k \times n + r$ where $0 \leq r < n$ is the remainder when m is divided by n.

If $r = 0$, then $\gcd(m, n) = n$

If $r > 0$, then $d | (m - k \times n)$ since $d | m$ and $d | n$. Note that $r = m - k \times n$, so $d | r$. We can check that $\gcd(n, r) = d$.

Repeat the process with n and r: $n = h \times r + s$ and so forth till we get the remainder 0. Note that $s < r < n$ and the process will terminate.

Example: Find $\gcd(107{,}907{,}800,\ 94{,}128{,}804)$ using Euclidean algorithm

$$107{,}907{,}800 = 1 \times 94{,}128{,}804 + 13{,}778{,}996 \tag{9.1}$$

$$94{,}128{,}804 = 6 \times 13{,}778{,}996 + 11{,}454{,}828 \tag{9.2}$$

$$13{,}778{,}996 = 1 \times 11{,}454{,}828 + 2{,}324{,}168 \tag{9.3}$$

$$11{,}454{,}828 = 4 \times 2{,}324{,}168 + 2{,}158{,}156 \tag{9.4}$$

$$2{,}324{,}168 = 71 \times 2{,}158{,}156 + 166{,}012 \tag{9.5}$$

$$2{,}158{,}156 = 13 \times 166{,}012 + 0$$

Hence, $\gcd(107{,}907{,}800, 94{,}128{,}804) = 166{,}012$ which agrees with what we found earlier. There is a useful result from the Euclidean algorithm. We can now write the $\gcd(m, n)$ as an integer combination of m and n by reversing the above process.

We wish to write $\gcd(m, n)$ as a linear integer combination of m and n.

We start from Equation (9.5):

$$
\begin{aligned}
166{,}012 &= 2{,}324{,}168 - 1 \times 2{,}158{,}156 \,(\text{using (4) gives}) \\
&= 2{,}324{,}168 - 1 \times (11{,}454{,}828 - 4 \times 2{,}324{,}156) \\
&= -11{,}454{,}828 + 5 \times 2{,}324{,}168 \,(\text{using (3) gives}) \\
&= -11{,}454{,}828 + 5 \times (13{,}778{,}996 - 1 \times 11{,}454{,}828) \\
&= 5 \times 13{,}778{,}996 - 6 \times 11{,}454{,}828 \,(\text{using (2) gives}) \\
&= 5 \times 13{,}778{,}996 - 6 \times (94{,}128{,}804 - 6 \times 12{,}778{,}996)
\end{aligned}
$$

$$= -6 \times 94{,}128{,}804 + 41 \times 13{,}778{,}996 \,(\text{using }(1)\text{ gives})$$
$$= -6 \times 94{,}128{,}804 + 41 \times (107{,}907{,}800 - 1 \times 94{,}128{,}804)$$
$$= 41 \times 107{,}907{,}800 - 47 \times 94{,}128{,}804$$

Hence $166{,}012 = \mathbf{41} \times 107{,}907{,}800 - \mathbf{47} \times 94{,}128{,}804$

Exercise

Determine $\gcd(152,\ 398)$ and write $\gcd(152,\ 398)$ as a combination of 152 and 398.

Solution:

Using the Euclidean algorithm,

$$398 = 2 \times 152 + 94$$
$$152 = 1 \times 94 + 58$$
$$94 = 1 \times 58 + 36$$
$$58 = 1 \times 36 + 22$$
$$36 = 1 \times 22 + 14$$
$$22 = 1 \times 14 + 8$$
$$14 = 1 \times 8 + 6$$
$$8 = 1 \times 6 + 2$$
$$6 = 3 \times 2 + 0$$

So $\gcd(152, 398) = 2$.

Reversing the above steps,

$$2 = 8 - 6 = 2 \times 8 - 14 = 2 \times 22 - 3 \times 14$$
$$= 5 \times 22 - 3 \times 36 = 5 \times 58 - 8 \times 36$$
$$= 13 \times 58 - 8 \times 94 = 13 \times 152 - 21 \times 94$$
$$= 55 \times 152 - 21 \times 398$$

Thus, $\gcd(152, 398) = \mathbf{55} \times 152 - \mathbf{21} \times 398$

9.1.5 *Modular Arithmetic Definitions and Operations*

9.1.5.1 *Definition: Modulo n*

Let $I = \{\ldots, -4, -3, -2, -1, 0, 1, 2, 3, 4, 5, \ldots\}$ be the set of integers. Let $n > 1$ be a positive integer. For two integers $a, b \in I$, we have $a \equiv b$ modulo n if n divides $a - b$, that is $n|(a - b)$ or n is a factor of $(a - b)$.

Suppose we take $n = 7$, then $-11 \equiv 3 \pmod 7$ since $7|(-11 - 3)$. Also, $-9 \equiv -2 \equiv 5 \equiv 12 \equiv 19 \equiv 26 \equiv 33 \pmod 7$

Schematic of wrap around is shown in Figure 9.1.

For integer $n > 1$, we define addition modulo n as follows:

For $a, b \in I$, $a + b \equiv r \pmod n$ where r is the remainder when the sum $a + b$ is divided by n. Note that $r \in \{0, 1, 2, \ldots, n - 1\}$. Sometimes we write $a + b = r \pmod n$.

We similarly define multiplication modulo n. For $a, b \in I, a \times b \equiv s \pmod n$ where s is the remainder when the product $a \times b$ is divided by n. So $s \in \{0, 1, 2, \ldots, n - 1\}$.

Exercise

Let $2 + 17 + 32 \equiv r \pmod 7$. What is r?

(a) 1 (mod 7)
(b) 2 (mod 7)
(c) 3 (mod 7)

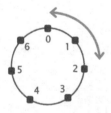

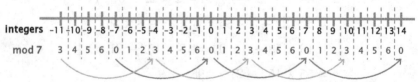

Figure 9.1: Illustration of schematic wraparound of modulo n.

Solution: Choice **b** is correct.

$$2 + 17 + 32 \equiv 2 + 3 + 4 \equiv 2 \ (\text{mod } 7)$$

9.1.5.2 *Additive and Multiplicative Identity*

Within the set of integers $I = \{\ldots, -2, -1, 0, 1, 2, 3, \ldots\}$ we have two special integers, namely **0** and **1**, with the following properties: for any integer $m \in I$,

- $m + \mathbf{0} = \mathbf{0} + m = m$
- $m \times \mathbf{1} = \mathbf{1} \times m = m$

0 is the additive identity and **1** is the multiplicative identity under the usual operations of addition and multiplication.

In modulo n arithmetic, we see that **0** and **1** are also the additive and multiplicative identities respectively.

For any integer m,

- $m + \mathbf{0} = \mathbf{0} + m = m (\text{mod } n)$, and
- $m \times \mathbf{1} = \mathbf{1} \times m = m (\text{mod } n)$

9.1.5.3 *Additive Inverse*

For the integers $I = \{\ldots, -2, -1, 0, 1, 2, 3, \ldots\}$, **0** is the additive identity. For each $m \in I$, we can find an integer u such that $m + u = u + m = \mathbf{0}$. Simply take $u = -m \in I$. We call $u = -m$ the additive inverse of m.

For example, -5 is the additive inverse of 5 since $5 + (-5) = \mathbf{0}$.
29 is the additive inverse of -29.

In modulo n arithmetic, **0** is the additive identity and again every integer m has an additive inverse, namely $-m \ (\text{mod } n)$.

$$m + (-m) = (-m) + m = \mathbf{0} \ (\text{mod } n).$$

Question: Is there anything different doing arithmetic in modulo n compared to integers?

9.1.5.4 *Multiplicative Inverse*

For the integers $I = \{\ldots, -2, -1, 0, 1, 2, 3, \ldots\}$, **1** is the multiplicative identity. Do we have multiplicative inverses?

For each $m \neq 0$, can we find an integer v such that $m \times v = v \times m = 1$?

Take $m = 5$. We have $5 \times (1/5) = 1$, but the problem is $1/5$ is *not* an integer.

Inside the set of integers $I = \{\ldots, -2, -1, 0, 1, 2, 3, \ldots\}$, no integer has a multiplicative inverse other than 1.

Trivially, 1 is its multiplicative inverse since $1 \times 1 = 1$.

Question: What about doing arithmetic, modulo n?

9.1.5.5   *Modulo 7: Multiplication Table*

Consider multiplication mod 7. With modulo 7, we just need to consider integers 0, 1, 2, 3, 4, 5, and 6 since every integer m is equivalent to one of them.

The multiplication table for mod 7 is shown in Figure 9.2 below. What do we see?

Again 1 is the multiplicative identity. For each $m \neq 0$, can we find v such that $m \times v = v \times m = 1 \pmod 7$?

From the table,

- $2 \times 4 = 4 \times 2 = 8 = 1 \pmod 7$
- $3 \times 5 = 5 \times 3 = 15 = 1 \pmod 7$
- $6 \times 6 = 36 = 1 \pmod 7$

So 2 and 4 are multiplicative inverses of each other, as are 3 and 5. Note that 6, as well as 1, are the multiplicative inverse of itself. So every integer other than 0 (obviously) has a multiplicative inverse under modulo 7.

×	0	1	2	3	4	5	6
0	0	0	0	0	0	0	0
1	0	1	2	3	4	5	6
2	0	2	4	6	1	3	5
3	0	3	6	2	5	1	4
4	0	4	1	5	2	6	3
5	0	5	3	1	6	4	2
6	0	6	5	4	3	2	1

$2 \times 6 \equiv 5 \pmod 7$

Figure 9.2:   Multiplication table for modulo 7.

9.1.6 *Why are Multiplicative Inverses Important?*

Suppose integer h has a multiplicative inverse denoted as h^{-1} (mod p).

Then $h \times u = h \times v \pmod{p}$ implies that $u = v \pmod{p}$. In other words, the cancellation law holds. To see this:

$$h \times u = h \times v \Rightarrow h^{-1} \times (h \times u) = h^{-1} \times (h \times v) \pmod{p}$$

$$\text{associative property}: (h^{-1} \times h) \times u = (h^{-1} \times h) \times v \pmod{p}$$

$$\text{so } u = 1 \times u = 1 \times v = v \pmod{p}$$

Similarly, $u \times h = v \times h \pmod{p}$ implies that $u = v \pmod{p}$

That each element has an inverse is a requirement of algebraic structures known as groups which have widespread applications in mathematics, physics, and cryptography.

Let us consider another example: arithmetic Modulo 6.

With modulo 6, we just need to consider the integers 0, 1, 2, 3, 4, and 5 since every integer m is equivalent to one of them. The multiplication table mod 6 is shown in Figure 9.3.

From the table, 2, 3, and 4 have no multiplicative inverse. 1 and 5 are their multiplicative inverses. $5 \times 5 = 25 = 1 \pmod{6}$.

We can check if there is a cancellation law for modulo 6.

From the table, we see that:

- $2 \times 2 = 2 \times 5 \pmod{6}$ but $2 \neq 5 \pmod{6}$
- $3 \times 3 = 3 \times 5 \pmod{6}$ but $3 \neq 5 \pmod{6}$
- $4 \times 2 = 4 \times 5 \pmod{6}$ but $2 \neq 5 \pmod{6}$
- $4 \times 1 = 4 \times 4 \pmod{6}$ but $1 \neq 4 \pmod{6}$

$2 \times 4 = 8 \equiv 2 \pmod{6}$

×	0	1	2	3	4	5
0	0	0	0	0	0	0
1	0	1	2	3	4	5
2	0	2	4	0	2	4
3	0	3	0	3	0	3
4	0	4	2	0	4	2
5	0	5	4	3	2	1

Figure 9.3: Multiplication table for modulo 6.

So the cancellation law does not hold in the above cases. However, 5 has a multiplicative inverse, namely itself, since $5 \times 5 = 1$ (mod 6). Checking the table, if $5 \times a = 5 \times b$ (mod 6), then $a = b$ (mod 6). That is the cancellation law holds for 5.

9.1.6.1   *Multiplicative Inverses in Modulo n*

Suppose $n = h \times k$ where h, $k > 1$ are integers. That is, n is not a prime. Such an integer is called a composite number. Can h have a multiplicative inverse mod n?

Suppose v is the multiplicative inverse of h. That is $v \times h = 1$ (mod n). Then

$$v \times n = v \times (h \times k) = (v \times h) \times k \equiv 1 \times k \equiv k \,(\mathrm{mod}\, n)$$

But $v \times n \equiv v \times 0 \equiv 0\,(\mathrm{mod}\, n)$, and so $k \equiv 0\,(\mathrm{mod}\, n)$, which is not possible since $n > k > 1$.

We see that h does not have a multiplicative inverse. The same applies to k. The conclusion is: if n is not a prime, there is some non-zero integer $m \neq 0$, such that m does not have a multiplicative inverse mod n.

We now check if the converse is true.

9.1.6.1.1   Multiplicative Inverses in Modulo n, n is a Prime

Suppose n is a prime. Let m be an integer with $1 < m < n$. Consider the products m^k where $k = 1, 2, 3, \ldots$

Since n is prime, n cannot divide m^k for all $k = 1, 2, 3, \ldots$.

If $n | m^k$ then $n | m$, which is not possible since $m < n$. That is, $m^k \neq 0$ (mod n) for all $k = 1, 2, 3, \ldots$.

Since m^k (mod n) can take on finitely many values, $1, 2, \ldots, n-1$, we must have $m^k = m^h$ (mod n) for some $k > h > 1$. That is $m^{k-h} = 1$ (mod n).

Now, $m \times m^{k-h-1} = m^{k-h} \equiv 1$ and m^{k-h-1} is the multiplicative inverse of m.

The conclusion is: if n is prime, then every integer $m \neq 0$, has a multiplicative inverse mod n.

Numerical Illustration

Take $n = 7$, a prime and take any non-zero number say 3. Consider $3, 3^2, 3^3, 3^4, \ldots \bmod 7$. As these powers of 3 can only take on finitely many values from 1, 2, 3, 4, 5, 6, there will be a repetition.

- $3^1 \equiv 3 \,(\bmod\, 7), 3^2 \equiv 2 \,(\bmod\, 7), 3^3 \equiv 6 \,(\bmod\, 7),$
- $3^4 \equiv 4 \,(\bmod\, 7), 3^5 \equiv 5 \,(\bmod\, 7), 3^6 \equiv 1 \,(\bmod\, 7),$
- $3^7 \equiv 3 \,(\bmod\, 7), 3^8 \equiv 2 \,(\bmod\, 7), 3^9 \equiv 6 \,(\bmod\, 7),$

So $3^1 \equiv 3^7 \equiv 3 \,(\bmod\, 7)$. Thus, $3^{7-1} \equiv 3^6 \equiv 1 \,(\bmod\, 7)$ as we can see in the calculation above. So the inverse of 3 is $3^5 \equiv 5 \,(\bmod\, 7)$. Check: $3 \times 5 = 5 \times 3 = 15 = 1 \,(\bmod\, 7)$.

9.1.6.2 *Polynomials with Coefficients Mod p*

A polynomial in variable x of degree $n \geq 0$ is an expression of the form

$$F(x) = a_n x^n + a_{n-1} x^{n-1} + \cdots + a_0$$

where the coefficients $a_n, a_{n-1}, \ldots, a_0$ are integers modulo p and $a_n \neq 0$.

Let Π_3 denote the set of polynomials in variable x with integer coefficients taken modulo 3. The following are elements in Π_3.

- $F_1(x) = 2x^5 + x^4 + 2 \times +1 \,(\bmod\, 3)$
- $F_2(x) = x^3 + 2x^2 + 2x + 2 \,(\bmod\, 3)$
- $F_3(x) = 2x^2 + x + 1 \,(\bmod\, 3)$
- $F_4(x) = x + 2 \,(\bmod\, 3)$
- $F_5(x) = 2 \,(\bmod\, 3)$

9.1.6.2.1 Polynomials mod $x^2 + 1$

The idea of doing arithmetic operations modulo n may be extended from integers to polynomials.

For $F(x), G(x) \in \Pi_3$, we define $F(x) \equiv G(x) \,(\bmod\, x^2 + 1)$ if $x^2 + 1$ divides $(F(x) - G(x))$, that is $(x^2 + 1)|(F(x) - G(x))$. Some examples are:

- $F_1(x) = 2x^5 + x^4 + 2x + 1$

$$= (x^2 + 1)(2x^3 + x^2 + x + 2) + x + 2$$

$$\equiv x + 2 \,(\bmod\, x^2 + 1)$$

- $F_2(x) = x^3 + 2x^2 + 2x + 2$

  $= (x^2 + 1)(x + 2) + x$

  $\equiv x (\operatorname{mod} x^2 + 1)$
- $F_3(x) = 2x^2 + x + 1$

  $= 2(x^2 + 1) + x + 2$

  $\equiv x + 2 \,(\operatorname{mod} x^2 + 1)$
- $F_4(x) = x + 2 \equiv F_1(x) \equiv F_3(x)(\operatorname{mod} x^2 + 1)$
- $F_5(x) = 2 \,(\operatorname{mod} x^2 + 1)$

What have we observed from the examples? Every $F(x) \in \Pi_3$ is equivalent to a polynomial $rx+s$, mod $x^2 + 1$ where $r, s \in \{0, 1, 2\}$. So there are 9 representative polynomials, three choices for r, and three choices for s (mod $x^2 + 1$). They are:

- $H_0(x) = 0, H_1(x) = 1, H_2(x) = 2,$
- $H_3(x) = x, H_4(x) = x + 1, H_5(x) = x + 2,$
- $H_6(x) = 2x, H_7(x) = 2x + 1, H_8(x) = 2x + 2,$

We can understand the structure of Π_3 by looking at the addition and multiplication of these nine representative polynomials.

Example

Consider $H_5(x) = x + 2$ and $H_7(x) = 2x + 1$

Sum: $H_5(x) + H_7(x) = (x + 2) + (2x + 1) = 3x + 3 \equiv 0 = H_0(x) \,(\operatorname{mod} x^2 + 1)$

$$\text{Product:} \quad H_5(x) \times H_7(x) = (x + 2)(2x + 1)$$
$$= 2x^2 + 5x + 2 \operatorname{mod} 3$$
$$= 2x^2 + 2x + 2 \operatorname{mod} 3$$
$$= 2(x^2 + 1) + 2x$$
$$= 2x = H_6(x) \quad \operatorname{mod} x^2 + 1$$

9.1.6.2.2 Polynomials: Addition mod $x^2 + 1$

Let $\oplus$ denote addition of polynomials in Π_3, mod $x^2 + 1$. For $F(x)$, $G(x) \in \Pi_3$, we define

$$F(x) \oplus G(x) = F(x) + G(x) \bmod x^2 + 1$$

Figure 9.4 provides the results of addition mod $x^2 + 1$.

Observe the pattern in the table, like Sudoku, each $H_i, i = 0, 1, \ldots, 8$ appears exactly once in each row and column.

9.1.6.2.3 Polynomials: Multiplication mod $x^2 + 1$

Let $\otimes$ denote multiplication of polynomials in Π_3, mod $x^2 + 1$. For $F(x)$, $G(x) \in \Pi_3$, we define $F(x) \otimes G(x) = F(x) \times G(x) \bmod x^2 + 1$.

Figure 9.5 provides the results of multiplication mod $x^2 + 1$.

From the tables, we see that H_0 and H_1 are the additives and multiplicative identities respectively. That is for all $F(x) \in \Pi_3$, we have

$$H_0 \oplus F(x) = F(x) \oplus H_0 = F(x) \bmod x^2 + 1, \text{ and}$$

$$H_1 \otimes F(x) = F(x) \otimes H_1 = F(x) \bmod x^2 + 1$$

From the tables, we can also see that every polynomial $F(x) \in \Pi_3$ in has an additive inverse, namely $-F(x)$. Also, every $F(x) \in \Pi_3$ has a multiplicative inverse $F^{-1}(x)$. H_3 and H_6 are inverses of each other as are H_4 and H_5, and H_7 and H_8. H_1 and H_2 are self-inverses.

$\oplus$	H_0	H_1	H_2	H_3	H_4	H_5	H_6	H_7	H_8
H_0	H_0	H_1	H_2	H_3	H_4	H_5	H_6	H_7	H_8
H_1	H_1	H_2	H_0	H_4	H_5	H_3	H_7	H_8	H_6
H_2	H_2	H_0	H_1	H_5	H_3	H_4	H_8	H_6	H_7
H_3	H_3	H_4	H_5	H_6	H_7	H_8	H_0	H_1	H_2
H_4	H_4	H_5	H_3	H_7	H_8	H_6	H_1	H_2	H_0
H_5	H_5	H_3	H_4	H_8	H_6	H_7	H_2	H_0	H_1
H_6	H_6	H_7	H_8	H_0	H_1	H_2	H_3	H_4	H_5
H_7	H_7	H_8	H_6	H_1	H_2	H_0	H_4	H_5	H_3
H_8	H_8	H_6	H_7	H_2	H_0	H_1	H_5	H_3	H_4

Figure 9.4: Results of polynomials: addition mod $x^2 + 1$.

⊗	H_0	H_1	H_2	H_3	H_4	H_5	H_6	H_7	H_8
H_0	H_0	H_0	H_0	H_0	H_0	H_0	H_0	H_0	H_0
H_1	H_0	H_1	H_2	H_3	H_4	H_5	H_6	H_7	H_8
H_2	H_0	H_2	H_1	H_6	H_8	H_7	H_3	H_5	H_4
H_3	H_0	H_3	H_6	H_2	H_5	H_8	H_1	H_4	H_7
H_4	H_0	H_4	H_8	H_5	H_6	H_1	H_7	H_2	H_3
H_5	H_0	H_5	H_7	H_8	H_1	H_3	H_4	H_6	H_2
H_6	H_0	H_6	H_3	H_1	H_7	H_4	H_2	H_8	H_5
H_7	H_0	H_7	H_5	H_4	H_2	H_6	H_8	H_3	H_1
H_8	H_0	H_8	H_4	H_7	H_3	H_2	H_5	H_1	H_6

Figure 9.5: Results of polynomials: multiplication mod $x^2 + 1$.

9.1.6.3 Do Multiplicative Inverses Always Exist Modulo a Polynomial?

With mod n, multiplicative inverses exist for all non-zero integers when n is prime. The same applies to arithmetic modulo a polynomial. The polynomial $x^2 + 1$ (mod 3) we chose in the earlier example is irreducible in that it cannot be factored into products of polynomials other than 1 and itself mod 3. That is $x^2 + 1 \neq (ax+b)(cx+d)$ (mod 3) for any choice of integers a, b, c, and d.

What if we take mod $x^2 + 2$?

Let us consider multiplication mod $x^2 + 2$. Again every polynomial in Π_3 is equivalent to one of $H_0, H_1, \ldots, H_8$ as before. The multiplication table mod $x^2 + 2$ is shown in Figure 9.6.

The non-zero polynomials H_4, H_5, H_7, and H_8 have no multiplicative inverse. The reason is because $x^2 + 2 = (x + 2)(x + 1)$ mod 3. That is, $x^2 + 2$ is a reducible polynomial. It can be factored as a product of two or more polynomials of a lower degree.

9.1.6.4 Why Do We Care About Arithmetic Modulo a Polynomial?

Groups and fields are algebraic structures that have widespread applications in computer science, cryptography, and fintech. One way to understand finite fields is to study polynomials with coefficients modulo p, where p is a prime. Much information about the field is

$\otimes$	H_0	H_1	H_2	H_3	H_4	H_5	H_6	H_7	H_8
H_0	H_0	H_0	H_0	H_0	H_0	H_0	H_0	H_0	H_0
H_1	H_0	H_1	H_2	H_3	H_4	H_5	H_6	H_7	H_8
H_2	H_0	H_2	H_1	H_6	H_8	H_7	H_3	H_5	H_4
H_3	H_0	H_3	H_6	H_1	H_4	H_7	H_2	H_5	H_8
H_4	H_0	H_4	H_8	H_4	H_8	H_0	H_8	H_0	H_4
H_5	H_0	H_5	H_7	H_7	H_0	H_5	H_5	H_7	H_0
H_6	H_0	H_6	H_3	H_2	H_8	H_5	H_1	H_7	H_4
H_7	H_0	H_7	H_5	H_5	H_0	H_7	H_7	H_8	H_0
H_8	H_0	H_8	H_4	H_8	H_4	H_0	H_4	H_0	H_8

Figure 9.6: Results of polynomials: multiplication mod $x^2 + 2$.

contained in the addition and multiplication tables, such as those we computed earlier. An effective way to understand abstract ideas is to work and practice with numerical computations.

Bibliography

Rosen, K. (2010). *Elementary Number Theory and its Applications*. 6th ed., Chap. 3. Pearson, New Jersey, NJ.
Stillwell, J. (1994). *Elements of Algebra*. Springer, New York, NY.
Tattersall, J. (2005). *Elementary Number Theory in Nine Chapters*. Cambridge University Press, Cambridge, UK, pp. 64–86.

9.2 Sample Questions

Please select the most appropriate response.

Question 1

Let x, y, and z be the smallest positive integers that satisfy the following equations:

$$71 \equiv x \,(\mathrm{mod}\,8), 78 + y \equiv 3 \,(\mathrm{mod}\,5) \text{ and } 89 \equiv (z + 3) \,(\mathrm{mod}\,4).$$

Which number among x, y, and z has the smallest value?

(a) x
(b) y
(c) z

Question 2

For multiplication modulo 12, which of the following statements about multiplicative inverses is correct?

(a) The elements 5, 7, and 11 have multiplicative inverses
(b) The elements 1, 3, 7, and 9 have multiplicative inverses
(c) The elements 3, 5, and 11 have multiplicative inverses

Question 3

What is the GCD of 43,832,250 and 85,387,500?

(a) 341,550
(b) 417,450
(c) 569,250

Question 4

What is the least common multiple of 43,832,250 and 85,387,500? The least common multiple of m and n, denoted at $\text{lcm}(m, n)$ is the smallest positive integer that is a multiple of m and also of n.

(a) 6,534,878,500
(b) 6,574,837,500
(c) 6,674,338,500

The following information applies to questions 5 to 7.

We consider polynomials in variable x with integer coefficients modulo 5.

Let $F(x) = x + 3$ and $G(x) = 2x + 1$.

Question 5

Which of the following is equal to $F(x) + 2G(x)$?

(a) $3x + 4$
(b) $4x + 2$
(c) 0

Question 6

Which of the following is equal to the product $F(x) \times G(x)$ modulo $(x^2 + 2)$?

(a) $4x + 2$
(b) $2x + 4$
(c) None of the above

Question 7

Which of the following is the multiplicative inverse of $F(x)$ modulo $(x^2 + 2)$?

(a) $4x + 3$
(b) $4x + 2$
(c) $2x + 4$

Solutions

Question 1

Solution: Option **c** is correct

(a) $71 \equiv x \pmod 8$ $71 = 64 + 7 \equiv 0 + 7 \pmod 8$ So $x \equiv 7 \pmod 8$. The smallest value for x is $x = 7$.

(b) $78 + y \equiv 3 \pmod 5$ $78 + y - 3 = 0 \pmod 5$ $75 + y = 0 \pmod 5$ $0 + y = 0 \pmod 5$.

Since y has to be positive, we pick $y = 5$ as it is the smallest positive integer.

(c) $89 \equiv (z + 3) \pmod 4$.

$89 - 3 = z \pmod 4$. $86 = z \pmod 4$ $2 = z \pmod 4$.

Hence the least positive value of z must be 2.

Question 2

Solution: Option **a** is correct

Note: $1 \times 1 = 1 \pmod{12}$, $5 \times 5 = 1 \pmod{12}$, $7 \times 7 = 1 \pmod{12}$, $11 \times 11 = 1 \pmod{12}$. The remaining elements 0, 2, 3, 4, 6, 8, 9 and 10 do not have multiplicative inverses. In mod n, a non-zero element m will have a multiplicative inverse if and only if $\gcd(m, n) = 1$.

Question 3

Solution: Option **c** is correct

We apply the Euclidean algorithm

$$85{,}387{,}500 = 1 \times 43{,}832{,}250 + 41{,}555{,}250$$

$$43{,}832{,}250 = 1 \times 41{,}555{,}250 + 2{,}277{,}000$$

$$41{,}555{,}250 = 18 \times 2{,}277{,}000 + 569{,}250$$

$$2{,}277{,}000 = 4 \times 569{,}250 + 0$$

Hence $\gcd(43{,}832{,}250, 85{,}387{,}500) = 569{,}250$

Question 4

Solution: Option **b** is correct

From the Euclidean algorithm,

$$2{,}277{,}000 = 4 \times 569{,}250$$

$$41{,}555{,}250 = 18 \times 2{,}277{,}000 + 569{,}250$$

$$= (18 \times 4 + 1) \times 569{,}250$$

$$43{,}832{,}250 = 1 \times 41{,}555{,}250 + 2{,}277{,}000$$

$$= (73 + 4) \times 569{,}250 = 77 \times 569{,}250$$

$$85{,}387{,}500 = 1 \times 43{,}832{,}250 + 41{,}555{,}250$$

$$= (77 + 73) \times 569{,}250 = 150 \times 569{,}250$$

We see that $43{,}832{,}250 = \mathbf{77} \times 569{,}250$ and $85{,}387{,}500 = \mathbf{150} \times 569{,}250$

Thus $\text{lcm}(43{,}832{,}250, 85{,}387{,}500) = \mathbf{77} \times \mathbf{150} \times 569{,}250 = 6{,}574{,}837{,}500$

Note that $\text{lcm}(m,n) \times gcd(m,n) = m \times n$.

Question 5

Solution: Option **c** is correct

$F(x) + 2G(x) = x + 3 + 2(2x + 1) = 5x + 5 = 0 \ (\text{mod } 5)$.

Question 6

Solution: Option **b** is correct

$F(x) \times G(x) = (x+3) \times (2x+1) = 2x^2 + 2x + 3 = 2x + 4 \bmod (x^2 + 2)$

Question 7

Solution: Option **a** is correct

Let $F^{-1}(x) = ax + b$. Then $(x + 3)(ax + b) = 1 \bmod (x^2 + 2)$.

Expanding the left-hand side, we have

$$ax^2 + (3a + b)x + 3b = 1 \bmod (x^2 + 2)$$

$$a(x^2 + 2) + (3a + b)x + 3b - 2a = 1 \bmod (x^2 + 2)$$

Thus $3a + b = 0$ and $3b - 2a = 1 \bmod 5$. Solving these two equations, we have $a = 4$ and $b = 3$. The multiplicative inverse of $F(x)$ is $F^{-1}(x) = 4x + 3$.

Note: Finding the multiplicative inverse involves solving a set of simultaneous equations. However to obtain the answer to this question, we can simply multiply each of the choices (a), (b), and (c) with $F(x)$ to see if the resulting answer is 1.

Chapter 10

Matrix Operations

10.1 Matrices

In mathematics, a matrix (plural matrices) is a rectangular *array* (see irregular matrix) of numbers, symbols, or expressions, arranged in *rows* and *columns*.

10.1.1 *Learning Objectives*

- Understand basic matrix operations and associated properties.

10.1.2 *Main Takeaways*

10.1.2.1 *Main Points*

- Addition and Subtraction: Matrices must have the same size (each matrix has the same number of rows and the same number of columns as the other) before they can be added or subtracted element by element. An m-by-n or $m \times n$ matrix has m rows and n columns.
- Multiplication: The product AB of two matrices A and B is defined if and only if the number of columns of the left matrix A is the same as the number of rows of the right matrix B. If A is an m-by-n matrix and B is an n-by-p matrix, then their matrix product AB is the m-by-p matrix whose i, j entry is given by the dot product of the corresponding ith row of A and the corresponding jth column of B.

- The transpose of a matrix is obtained by interchanging its rows and columns. If C is a 2×3 matrix, its transpose C^t will be a 3×2 matrix.

10.1.2.2   *Main Terms*

- **Transpose:** In linear algebra, the transpose of a matrix is an operator which flips a matrix over its diagonal; that is, it switches the row and column indices of matrix A by producing another matrix, often denoted by A^t.
- **Associative property:** The associative property states that you can add or multiply regardless of how the numbers or matrices are grouped. Let A, B, and C be matrices of the same size. The sums $(A + B + C = A + (B + C) = A + B + C$ are all equal. Let D, E, and F be matrices of sizes $p \times q$, $q \times r$, and $r \times s$, respectively. The products $(D \times E) \times F = D \times (E \times F) = D \times E \times F$ are all equal with size $p \times s$.
- **Square Matrix:** In mathematics, a square matrix is a matrix with the same number of rows and columns. An n-by-n matrix is known as a square matrix of order n.
- **Minimum Variance Portfolio:** A collection of securities that minimizes the price volatility of the overall portfolio.
- **Lagrange Multiplier:** The process to find the maximum and minimum values of a function of various variables on occasions where the variables are restricted by additional constraints.

The study of matrices, which are rectangular arrays of numbers, has yielded many applications within and outside of mathematics. In this section, we review the basic matrix operations and associated properties before illustrating their use with some examples.

Matrices enable one to have compact descriptions of models as well as efficient computations. Often the results from linear algebra can offer interesting insights into the model or problem.

10.1.3   *Matrix Operations*

Matrices are rectangular arrays of numbers arranged in rows and columns as shown in Figure 10.1.

This is a 2×3 matrix with 2 rows and 3 columns.

Element in the 1st row and 1st column = 6

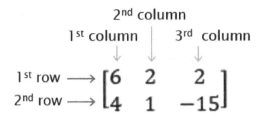

2nd column

1st column | 3rd column

1st row $\longrightarrow$ $\begin{bmatrix} 6 & 2 & 2 \\ 4 & 1 & -15 \end{bmatrix}$
2nd row $\longrightarrow$

Figure 10.1: Introductory matrix.

Element in the 2nd row and 3rd column $= -15$

For a general $m \times n$ matrix, $A = (a_{ij})_{m \times n}$, a_{ij} denotes the element in the ith row and j^{th} column of A

$$A = \begin{pmatrix} a_{11} & \cdots & a_{1n} \\ \vdots & \ddots & \vdots \\ a_{m1} & \cdots & a_{mn} \end{pmatrix}.$$

We can add two matrices together only when they are the same size, that is, the same number of rows and columns.

Example

Let $A = \begin{pmatrix} 5 & 2 \\ 3 & 1 \end{pmatrix}$, $B = \begin{pmatrix} 6 \\ 3 \\ -5 \end{pmatrix}$, $C = \begin{bmatrix} 6 & 2 & 2 \\ 4 & 1 & -15 \end{bmatrix}$, $D = \begin{bmatrix} -4 & 3 & 0 \\ 3 & 2 & 8 \end{bmatrix}$.

We see that A is a 2×2 matrix, B is 3×1, C is 2×3 and so is D. The sum $C+D$ is allowed but not $A+B$ or $A+C$ or $B+C$.

Matrix addition is done element-wise. If $A = (a_{ij})_{m \times n}$ and $B = (b_{ij})_{m \times n}$, the sum $S = A+B = (s_{ij})_{m \times n}$ is such that $s_{ij} = a_{ij} + b_{ij}$ for all $1 \leq i, j \leq n$.

Example

The sum $C+D$ is given by

$$C+D = \begin{bmatrix} 6 & 2 & 2 \\ 4 & 1 & -15 \end{bmatrix} + \begin{bmatrix} -4 & 3 & 0 \\ 3 & 2 & 8 \end{bmatrix}$$

$$= \begin{bmatrix} 6-4 & 2+3 & 2+0 \\ 4+3 & 1+2 & -15+8 \end{bmatrix} = \begin{bmatrix} 2 & 5 & 2 \\ 7 & 3 & -7 \end{bmatrix}.$$

We note that $C+D = D+C$. So matrix addition is commutative.

10.1.4 *Transpose of a Matrix*

The transpose of a matrix is obtained by interchanging its rows and columns. If C is a 2×3 matrix, its transpose C^t will be a 3×2 matrix.

Examples

The transpose of $C = \begin{bmatrix} 6 & 2 & 2 \\ 4 & 1 & -15 \end{bmatrix}$ is $C^t = \begin{bmatrix} 6 & 4 \\ 2 & 1 \\ 2 & -15 \end{bmatrix}$.

The transpose of $A = \begin{pmatrix} 5 & -2 \\ 3 & 1 \end{pmatrix}$ is $A^t = \begin{pmatrix} 5 & 3 \\ -2 & 1 \end{pmatrix}$.

The transpose of $B = \begin{bmatrix} 6 \\ 3 \\ -5 \end{bmatrix}$ is $B^t = \begin{bmatrix} 6 & 3 & -5 \end{bmatrix}$.

10.1.5 *Matrix Multiplication*

For two matrices, G (size $r \times s$) and H (size $u \times v$) the product $G \times H$ are defined if and only if $s = u$, that is the number of columns in G equals the number of rows in H. If $s = u$, then the product $G \times H$ is a matrix of size $r \times v$.

For example, with

$$A = \begin{pmatrix} 5 & -2 \\ 3 & 1 \end{pmatrix}, \quad B = \begin{pmatrix} 6 \\ 3 \\ -5 \end{pmatrix},$$

$$C = \begin{bmatrix} 6 & 2 & 2 \\ 4 & 1 & -15 \end{bmatrix}, \quad D = \begin{bmatrix} -4 & 3 & 0 \\ 3 & 2 & 8 \end{bmatrix}.$$

Permitted or defined: $A \times A$, $A \times C$, $A \times D$, $C \times B$, $D \times B$, $C \times D^t$, etc.

Note that $C \times A$ is not permitted or defined although $A \times C$ is. So matrix multiplication is not commutative. That is, $A \times B \neq B \times A$ in general.

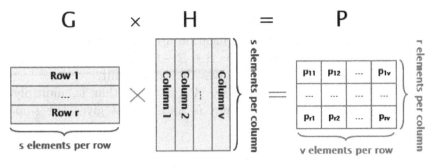

Figure 10.2: Graphical illustration of matrix multiplications.

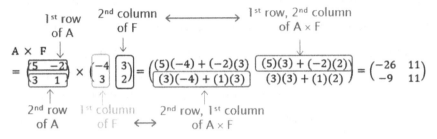

Figure 10.3: Graphical illustration on matrix multiplications.

10.1.5.1 *Matrix Multiplication: Mechanics as Illustrated Using Figure 10.2*

Let $G = (g_{ij})_{r \times s}$ and $H = (h_{ij})_{s \times v}$. Then the product $G \times H = P = (p_{ij})_{r \times v}$ is a $r \times v$ matrix whose ij element p_{ij} is given by

$$p_{ij} = g_{i1}h_{1j} + g_{i2}h_{2j} + g_{i3}h_{3j} + \cdots + g_{is}h_{sj} = \sum_{k=1}^{s} g_{ik}h_{kj}.$$

For the 2×2 matrices, A and F shown in Figure 10.3, we can form the product $A \times F$ since the number of columns of A equals the number of rows in F. The result is a 2×2 matrix.

Let $A = \begin{pmatrix} 5 & -2 \\ 3 & 1 \end{pmatrix}$ and $F = \begin{pmatrix} -4 & 3 \\ 3 & 2 \end{pmatrix}$. The working for computing $A \times F$ is shown below.

For example, to obtain the 1,2 entry in the first row, second column of $A \times F$, we perform the inner product of the first row of A with the second column of F, and so forth.

10.1.5.2 *Associative Property*

Both matrix addition and multiplication are associative in that
$(A + B) + C = A + (B + C) = A + B + C$, that is, adding A and
B first or adding B and C first gives the same result.

$(A \times B) \times C = A \times (B \times C) = A \times B \times C$, that is, multiplying A and B first or multiplying B and C first gives the same
result.

10.1.6   *Square Matrices: The Identity Matrix*

Consider the 2×2 matrix, $I_2 = \begin{pmatrix} 1 & 0 \\ 0 & 1 \end{pmatrix}$. For any 2×2 matrix $A = \begin{pmatrix} a & b \\ c & d \end{pmatrix}$, it is easy to check that $I_2 \times A = A \times I_2 = A = \begin{pmatrix} a & b \\ c & d \end{pmatrix}$.

The matrix $I_2 = \begin{pmatrix} 1 & 0 \\ 0 & 1 \end{pmatrix}$ is known as the 2×2 identity
matrix. For $n \times n$ matrices, the identity matrix is the $n \times n$
matrix with 1's along the main diagonal and 0 elsewhere. For
example, the 3×3 identity matrix is $I_3 = \begin{pmatrix} 1 & 0 & 0 \\ 0 & 1 & 0 \\ 0 & 0 & 1 \end{pmatrix}$ and more
generally

$$I_n = \begin{pmatrix} 1 & \cdots & 0 \\ \vdots & \ddots & \vdots \\ 0 & \cdots & 1 \end{pmatrix}.$$

10.1.6.1 *Invertible and Singular Matrices*

For an $n \times n$ matrix A, if there exists another $n \times n$ matrix B such
that $A \times B = B \times A = I_n$ the $n \times n$ identity matrix, we say that A
is invertible. The matrix B is called the inverse of A and is usually
written as $B = A^{-1}$. If A^{-1} does not exist, we say that A is singular
or it is not invertible.

Consider $A = \begin{pmatrix} 5 & -2 \\ 3 & 1 \end{pmatrix}$. We check that the matrix $B = \frac{1}{11}\begin{pmatrix} 1 & 2 \\ -3 & 5 \end{pmatrix}$
satisfies

$B \times A = A \times B = I_2 = \begin{pmatrix} 1 & 0 \\ 0 & 1 \end{pmatrix}$. Thus A is invertible and its inverse

is $A^{-1} = \frac{1}{11}\begin{pmatrix} 1 & 2 \\ -3 & 5 \end{pmatrix}$.

10.1.7 *Determinant*

For a 2×2 matrix, $A = \begin{pmatrix} a & b \\ c & d \end{pmatrix}$ we define its determinant as

$$\det(A) = \begin{vmatrix} a & b \\ c & d \end{vmatrix} = ad - bc.$$

If $\det(A) \neq 0$, then A is invertible and its inverse matrix is given by

$$A^{-1} = \frac{1}{ad - bc} \begin{pmatrix} d & -b \\ -c & a \end{pmatrix}.$$

This result holds generally for an $n \times n$ matrix. That is, a square $n \times n$ matrix M is invertible if and only if its determinant $\det(M) \neq 0$.

10.2 Why Do We Study Matrices?

Matrices have numerous applications in science and engineering. An electronic picture comprises pixels and we can use numbers to represent the color and light intensity of each pixel. Each picture can be thought of as a giant matrix, say M with size $10{,}000 \times 10{,}000(100{,}000{,}000$ numbers).

In the 1970s, when National Aeronautical and Space Administration (NASA) did space explorations, they wanted to send back as many pictures as possible. They used a computer on board the spacecraft to do a singular value decomposition to factorize M as $M = PVQ^t$ where V is a diagonal matrix of singular values. It turned out that for many pictures, only a few singular values were large and the rest can be ignored. So they could send back about 100,000 numbers instead of 100,000,000 per picture. Matrix multiplication on earth then allowed them to reconstruct the picture.

10.2.1 *Application in Finance: Risk and Expected Return Profiles of Investment Portfolios*

Suppose there are two assets A and B with the following risk-expected return characteristics:

Asset A: expected return $E(R_A) = 12\%$; standard deviation $\sigma_A = 20\%$

Asset B: expected return $E(R_B) = 15\%$; standard deviation $\sigma_B = 25\%$

The correlation between the returns on A and B is $\rho = 0.5$.

A way to use the information is to regard the return of asset A as a random variable R_A whose distribution has mean $E(R_A) = 12\%$ and standard deviation $\sigma_A = 20\%$. There is no need to make a further assumption about the return distribution.

A portfolio, P is $w_A = 40\%$ invested in A and $w_B = 60\%$ invested in B. What are the expected return and risks of the portfolio? The risk of an asset or portfolio is the standard deviation of its return distribution. The higher the standard deviation, the riskier the asset is since its return varies over a wide range with higher probability. In finance, the standard deviation of the return distribution is known as volatility.

The return of portfolio P, R_P is a combination of the returns of A and B, weighted by the amount invested in them, namely $R_P = w_A R_A + w_B R_B$.

The expected return of P:

$$E(R_P) = w_A \times E(R_A) + w_B \times E(R_B) = 0.4 \times 12\% + 0.6 \times 15\% = 13.8\%.$$

What about the risk of P? To find the standard deviation (risk), we first compute the variance.

$$\mathrm{Var}(R_P) = \mathrm{Var}(w_A \times R_A + w_B \times R_B) = w_A^2 \times \sigma_A^2 + w_B^2$$
$$\times \sigma_B^2 + 2 \times w_A \times w_B \times \rho \times \sigma_A \times \sigma_B = 0.0409$$

Note that correlation $\rho = \mathrm{Cov}(R_A, R_B)/\sigma_A \times \sigma_B$.

The standard deviation of the return of portfolio P, $\sigma_P = \sqrt{0.0409} = 20.22\%$.

Matrices allow us to do these computations compactly.

The variance–covariance matrix shows the variances down the diagonal and covariances for off-diagonal elements. For the two assets, it is

$$V = \begin{pmatrix} \sigma_A^2 & \mathrm{Cov}(R_A, R_B) \\ \mathrm{Cov}(R_A, R_B) & \sigma_B^2 \end{pmatrix} = \begin{pmatrix} 4\% & 2.5\% \\ 2.5\% & 6.25\% \end{pmatrix} \begin{matrix} \leftarrow \text{asset } A \\ \leftarrow \text{asset } B \end{matrix}$$

Note that $\sigma_A^2 = (20\%)^2 = 0.04 = 4\%$, $\sigma_B^2 = (25\%)^2 = 0.0625 = 6.25\%$,

$$\text{Cov}(R_A, R_B) = \rho \times \sigma_A \times \sigma_B = 0.5 \times 20\% \times 25\% = 2.5\%.$$

The return vector or matrix $R = \begin{pmatrix} E(R_A) \\ E(R_B) \end{pmatrix} = \begin{pmatrix} 12\% \\ 15\% \end{pmatrix} \begin{matrix} \leftarrow \text{ asset } A \\ \leftarrow \text{ asset } B \end{matrix}$.

The weight vector or matrix $w = \begin{pmatrix} w_A \\ w_B \end{pmatrix} = \begin{pmatrix} 40\% \\ 60\% \end{pmatrix} \begin{matrix} \leftarrow \text{ asset } A \\ \leftarrow \text{ asset } B \end{matrix}$.

The expected return of portfolio P is $E(R_P) = w^t R = \begin{pmatrix} 40\% & 60\% \end{pmatrix} \begin{pmatrix} 12\% \\ 15\% \end{pmatrix} = 13.8\%$.

If σ_P denotes the volatility (standard deviation) of R_P, then

$$\sigma_P^2 = w^t V w = \begin{pmatrix} 40\% & 60\% \end{pmatrix} \begin{pmatrix} 4\% & 2.5\% \\ 2.5\% & 6.25\% \end{pmatrix} \begin{pmatrix} 40\% \\ 60\% \end{pmatrix} = 4.09\%.$$

The standard deviation of return of portfolio P, $\sigma_P = \sqrt{4.09\%} = 20.22\%$.

10.2.1.1 *Impact of Different Correlation*

Figure 10.4 shows how the volatility of the portfolio return changes with differing weights of asset A and asset B in the portfolio as well as different correlations between the returns on asset A and on asset B.

			A	B				
		Return	12%	15%				
		Risk (vol)	20%	25%				

Weight of A	Weight of B	Return of Portfolio	Volatility (risk) of portfolio					
			$\rho = -0.3$	$\rho = 0.0$	$\rho = 0.3$	$\rho = 0.5$	$\rho = 0.7$	$\rho = 1$
0%	100%	15.00%	25.00%	25.00%	25.00%	25.00%	25.00%	25.00%
10%	90%	14.70%	21.98%	22.59%	23.18%	23.56%	23.94%	24.50%
20%	80%	14.40%	19.18%	20.40%	21.54%	22.27%	22.98%	24.00%
30%	70%	14.10%	16.71%	18.50%	20.13%	21.15%	22.12%	23.50%
40%	60%	13.80%	14.73%	17.00%	19.00%	20.22%	21.38%	23.00%
50%	50%	13.50%	13.46%	16.01%	18.20%	19.53%	20.77%	22.50%
60%	40%	13.20%	13.11%	15.62%	17.78%	19.08%	20.30%	22.00%
70%	30%	12.90%	13.76%	15.88%	17.76%	18.90%	19.98%	21.50%
80%	20%	12.60%	15.26%	16.76%	18.14%	19.00%	19.82%	21.00%
90%	10%	12.30%	17.41%	18.17%	18.90%	19.37%	19.83%	20.50%
100%	0%	12.00%	20.00%	20.00%	20.00%	20.00%	20.00%	20.00%

Figure 10.4: Relationship between the volatility of portfolio return and different weights as well as different correlations between the returns.

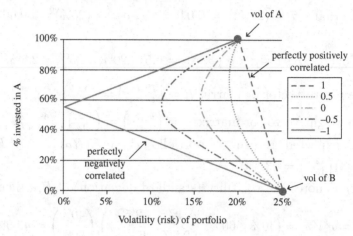

Figure 10.5: An illustration of the effect of correlation on portfolio volatility.

The effect of correlation is shown by plotting the percentage of the portfolio invested in asset A against the volatility of the portfolio return for different values of the correlation between the returns on asset A and asset B. The improved trade-off between return and volatility is obtained as the correlation between the returns on asset A and asset B decreases from 1 to -1. This relation is illustrated in Figure 10.5.

10.2.2 *Example with Five Assets*

The expected return, volatility, and correlation are as follows:

Assets	Expected Return Over Cash	Expected Volatility
A	11%	25%
B	9%	24%
C	10%	18%
D	5%	11%
E	4%	8%

Correlations				
Λ	B	C	D	E
1	0.5	0.5	0.4	0.2
0.5	1	0.4	0.2	0.2
0.5	0.4	1	0.5	0.3
0.4	0.2	0.5	1	0.5
0.2	0.2	0.3	0.5	1

Solution

The computed variance-covariance matrix is:

$$V = \begin{bmatrix} 6.25\% & 3.00\% & 2.25\% & 1.10\% & 0.40\% \\ 3.00\% & 5.76\% & 1.73\% & 0.53\% & 0.38\% \\ 2.25\% & 1.73\% & 3.24\% & 0.99\% & 0.43\% \\ 1.10\% & 0.53\% & 0.99\% & 1.21\% & 0.44\% \\ 0.40\% & 0.38\% & 0.43\% & 0.44\% & 0.64\% \end{bmatrix}$$

Its inverse is:

$$V^{-1} = \begin{bmatrix} 26.24 & -9.93 & -9.44 & -13.76 & 5.39 \\ -9.93 & 24.64 & -7.36 & 7.49 & -8.75 \\ -9.44 & -7.36 & 50.46 & -27.81 & -4.62 \\ -13.76 & 7.49 & -27.81 & 141.76 & -74.58 \\ 5.39 & -8.75 & -4.62 & -74.58 & 212.53 \end{bmatrix}$$

10.2.2.1 *Minimum Variance Portfolio*

How do we find the portfolio of the five assets with the lowest variance? This would require the solution of a minimization problem:

$$\min_w \sigma_P^2 = \min_w w^t V w \text{ subject to } w^t \mathbf{1} = 1$$

where w is the vector of weights of the assets in the portfolio and $\mathbf{1}$ denotes the 5×1 vector whose entries are all 1s.

This is a constrained optimization problem that is solved using the Lagrange method.

The first-order condition is:

$$\frac{\partial}{\partial w}(w^t V w - \lambda(w^t \mathbf{1} - 1)) = 2Vw + \lambda \mathbf{1} = 0$$

where λ is the Lagrange multiplier.

This first-order condition together with the constraint $w^t \mathbf{1} - 1 = 0$ allows us to obtain

$$w = -\frac{1}{2}\lambda V^{-1}\mathbf{1} \quad \text{and} \quad \lambda = -2\frac{1}{\mathbf{1}^t V^{-1}\mathbf{1}}.$$

Finally, eliminating λ gives

$$W_{\min} = \frac{V^{-1}\mathbf{1}}{\mathbf{1}^t V^{-1}\mathbf{1}}.$$

The minimum variance portfolio has the weights given by:

$$W_{\min} = \frac{V^{-1}\mathbf{1}}{\mathbf{1}^t V^{-1}\mathbf{1}} = \begin{bmatrix} -0.89\% \\ 3.60\% \\ 0.72\% \\ 19.60\% \\ 76.96\% \end{bmatrix} \begin{matrix} \leftarrow A \\ \leftarrow B \\ \leftarrow C \\ \leftarrow D \\ \leftarrow E \end{matrix}$$

Further, the minimum variance portfolio has a volatility of 7.70% and a return (over cash or the risk-free rate) of 4.36%. We also note that the portfolio is slightly short asset A.

10.2.2.2 *The Sharpe Ratio*

The Sharpe ratio is the ratio of the return (over cash) you get per unit of risk taken.

The risk-free rate R_{free} is what you can earn with no risk. The yield on a treasury bill or government bond is used as a proxy for the risk-free rate.

The Sharpe ratio is defined as:

$$\text{Sharpe Ratio} = \frac{R_P - R_{\text{free}}}{\sigma_P}$$

where R_P is the return on portfolio P and σ_P is the standard deviation of the return on portfolio P. Figure 10.6 depicts the Sharpe ratio as the slope of the line joining point P and R_{free}.

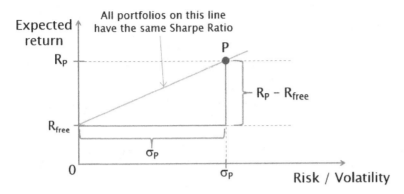

Figure 10.6: Graphical illustration of Sharpe ratio.

With five assets, A, B, C, D, and E, which portfolio of these assets have the highest Sharpe ratio? It turns out that the highest (maximum) Sharpe ratio portfolio has weights that satisfy

$$w_{\text{max Sharpe}} \propto \frac{V^{-1}\,R}{R^{t}\,V^{-1}\,R}$$

where $R = \begin{bmatrix} 11\% \\ 9\% \\ 10\% \\ 5\% \\ 4\% \end{bmatrix}$ is the vector of returns (over cash) for the various

assets.

Normalizing the vector, the highest Sharpe ratio portfolio has weights given by

$$W_{\text{max Sharpe}} = \begin{bmatrix} 7.83\% \\ 5.61\% \\ 24.05\% \\ 6.58\% \\ 55.93\% \end{bmatrix} \begin{matrix} \leftarrow A \\ \leftarrow B \\ \leftarrow C \\ \leftarrow D \\ \leftarrow E \end{matrix}$$

The Max Sharpe ratio portfolio has a volatility of 9.28% and a return (over cash) of 6.34%. This portfolio has no short positions.

How do we obtain all portfolios that provide the best available expected return and risk trade-off? These optimal portfolios are called efficient portfolios.

The efficient portfolios are the solutions to an optimization problem: maximize return for a given level of risk. Now if you find two

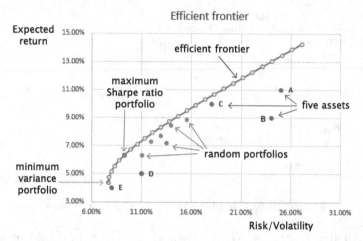

Figure 10.7:   Graphical illustration of the efficient frontier.

efficient portfolios, you can find all of them as the solution space is "two dimensional".

Now, have we found two efficient portfolios? One would be the minimum variance portfolio. The second intuitively would be the maximum Sharpe ratio portfolio and this portfolio is efficient. Hence, any linear combination of these two portfolios will be efficient. This will be a portfolio with weights: $w = \theta\, w_{\min} + (1-\theta) w_{\max \text{Sharpe}}$ where $w_{\min}$ is the vector of weights of the assets in the minimum variance portfolio and $w_{\max \text{Sharpe}}$ is the vector of weights of the assets in the maximum Sharpe ratio portfolio, and θ is any real number.

The efficient frontier is plotted in Figure 10.7 by varying the value of θ. The minimum variance portfolio has a value $\theta = 1$ and the maximum Sharpe ratio portfolio has $\theta = 0$. Note that θ can be negative or larger than 1.

Bibliography

Benninga, S. (2014). *Financial Modelling*. Chaps. 8 and 9. MIT Press, Cambridge, MA.

Fuller, L. (2017). *Basic Matrix Theory*. Chaps. 1, 2, and 6. Dover Books, New York, NY.

Bodie, Z., Kane, A., and Marcus, A. J. (2020). *Investments*. 12th ed., Chaps. 7 and 8. McGraw-Hill Irwin, New York, NY.

Elton, E. J., Gruber, M., Brown, S. J., and Goetzmann, W. N. (2013). *Modern Portfolio Theory and Investment Analysis.* 9th ed., Chaps. 5 and 6. Wiley, New York, NY.

10.3 Sample Questions

Please select the most appropriate response.

Question 1

If $A = \begin{bmatrix} 5 & 3 \\ 1 & 1 \end{bmatrix}$, then A^{-1} is equal to:

(a) $0.5 \begin{bmatrix} 1 & -3 \\ -1 & 5 \end{bmatrix}$

(b) $0.5 \begin{bmatrix} 1 & 3 \\ 1 & 5 \end{bmatrix}$

(c) $0.5 \begin{bmatrix} 5 & -3 \\ -1 & 1 \end{bmatrix}$

Question 2

If $A = \begin{bmatrix} 1 & -2 & 1 \\ 2 & 1 & 3 \end{bmatrix}$ and $B = \begin{bmatrix} 2 & 1 \\ 3 & 2 \\ 1 & 1 \end{bmatrix}$, then $A \times B$ is equal to

(a) $\begin{bmatrix} 3 & -2 \\ 10 & 7 \end{bmatrix}$

(b) $\begin{bmatrix} -3 & -2 \\ 10 & -7 \end{bmatrix}$

(c) $\begin{bmatrix} -3 & -2 \\ 10 & 7 \end{bmatrix}$

The following information applies to questions 3 to 5.

Let $A = \begin{bmatrix} 1 & -5 & 2 \end{bmatrix}$ and $B = \begin{bmatrix} 1 & -5 & 2 \\ 3 & 0 & -7 \end{bmatrix}$

Question 3

What is the result of the product AA^t?

(a) $[-20]$
(b) $[25]$
(c) $[30]$

Question 4

What is the result of the product BB^t?

(a) $\begin{bmatrix} 30 & -11 \\ 9 & 58 \end{bmatrix}$

(b) $\begin{bmatrix} 30 & -11 \\ -11 & 58 \end{bmatrix}$

(c) $\begin{bmatrix} 6 & -11 \\ -12 & 49 \end{bmatrix}$

Question 5

Let $U = \begin{bmatrix} x & y \end{bmatrix}$ be any 1×2 vector other than the zero vector $\mathbf{0} = \begin{bmatrix} 0 & 0 \end{bmatrix}$. What can we say about the product $U(BB^t)U^t$?

(a) We cannot conclude anything unless we know the numbers x and y.

(b) It is a 1×1 matrix (a number) and it may take on positive or negative values except 0.

(c) It is a 1×1 matrix (a number) and it is always a positive number.

Solutions

Question 1

Solution: Option **a** is correct.

Recall that the inverse of a 2×2 matrix is given by $\begin{bmatrix} a & b \\ c & d \end{bmatrix}^{-1} =$ $\frac{1}{ad-bc}\begin{bmatrix} d & -b \\ -c & a \end{bmatrix}$. The determinant of A is $5 \times 1 - 3 \times 1 = 2$ and the inverse of $A = \begin{bmatrix} 5 & 3 \\ 1 & 1 \end{bmatrix}$ is the matrix $0.5\begin{bmatrix} 1 & -3 \\ -1 & 5 \end{bmatrix}$.

Question 2

Solution: Option **c** is correct.

$$A \times B = \begin{bmatrix} 1 & -2 & 1 \\ 2 & 1 & 3 \end{bmatrix} \times \begin{bmatrix} 2 & 1 \\ 3 & 2 \\ 1 & 1 \end{bmatrix}$$

$$= \begin{bmatrix} 1 \times 2 + (-2) \times 3 + 1 \times 1 & 1 \times 1 + (-2) \times 2 + 1 \times 1 \\ 2 \times 2 + 1 \times 3 + 3 \times 1 & 2 \times 1 + 1 \times 2 + 3 \times 1 \end{bmatrix}$$

$$= \begin{bmatrix} -3 & -2 \\ 10 & 7 \end{bmatrix}$$

which is a 2×2 matrix.

Question 3

Solution: Option **c** is correct.

$$AA^t = \begin{bmatrix} 1 & -5 & 2 \end{bmatrix} \begin{bmatrix} 1 \\ -5 \\ 2 \end{bmatrix} = [1 \times 1 + (-5) \times (-5) + 2 \times 2] = [30].$$

Notice that if we consider A as a vector in the usual Cartesian coordinate framework, then AA^t is the square of the length of vector A. The quantity $AA^t > 0$ is always positive as long as $A \neq \mathbf{0}$ is not the zero vector $\mathbf{0} = \begin{bmatrix} 0 & 0 & 0 \end{bmatrix}$. We usually regard a 1×1 matrix such as AA^t as a number and talk about it being positive or negative, etc.

Notice that for any $1 \times n$ vector V, the product VV^t is a 1×1 matrix which is always a positive number whenever $V \neq \mathbf{0}$ is not equal to the zero-vector $\mathbf{0}$. Also the product V^tV is a $n \times n$ matrix.

Question 4

Solution: Option **b** is correct.

Question 5

Solution: Option **c** is correct.

Notice that UB is a 1×3 vector (matrix) and it is easy to see that $UB \neq \mathbf{0}$ since U is not the zero vector. We have $U(BB^t)U^t = (UB)(B^tU^t) = (UB)(UB)^t > 0$ as it is the square of the length of the non-zero 1×3 vector UB.

Chapter 11

Clustering

11.1 Clustering

This section reviews several hierarchical clustering methods as well as K-means clustering. These methods provide the foundation for more advanced clustering approaches. Clustering is often utilized in market segmentation and credit scoring. Clustering involves grouping objects or members that are similar based on a pre-defined set of characteristics.

11.1.1 *Learning Objectives*

- Discuss clustering analysis concepts, cluster validation, proximity measures, and limitations of clustering.
- Conduct cluster analysis using hierarchical clustering methods and the K-means method.

11.1.2 *Main Takeaways*

11.1.2.1 *Main Points*

- Hierarchical clustering methods initially classify the objects as singleton clusters (i.e., in an agglomerative manner). Objects are grouped iteratively until they become one cluster.
- The K-means method involves partitioning, which means the number of clusters is set before the clustering process is started.

Objects are moved among sets of clusters until the desired set is obtained.

- Different clustering algorithms can have different clustering solutions. This is because when applied to data, at each stage of the iterative process, different coefficients are used. These coefficients may be reflective of the computation of the proximity between the clusters and objects.
- Clustering is often used to identify common features in objects to conduct analysis and make decisions regarding the groups and their members. The subsequent analysis is made simpler when observations are grouped, as it becomes instinctive to spot outliers and reduce the original set's dimensionality.

11.1.2.2 *Main Terms*

- **Clustering:** The process organizes objects into groups or clusters based on predetermined characteristics or attributes (or clustering variables or criteria) to provide for more meaningful analysis.
- **Hierarchical clustering:** The process classifies objects as singleton clusters (i.e., in an agglomerative manner). Objects are grouped iteratively until they become one cluster.
- **Partitioning clustering:** In partitioning clustering, before clustering is performed, the number of clusters needs to be decided first.
- **Ward's linkage:** Minimizes the total within-cluster sum of squares in hierarchical clustering analysis. Clusters are formed incrementally and the solution with the smallest ESS, or error sum of squares, is the final solution.
- **Centroid linkage:** Formally defined using the squared Euclidean distance. The centroid is a vector that is the point of arithmetic mean position of all objects in a particular cluster.
- **Single-linkage:** The proximity (or similarity) of two clusters in the single-linkage algorithm is shown by the minimum distance calculated between any two objects out of all the possible combinations of objects in the clusters.
- **Complete-linkage:** The proximity of the two clusters in the complete-linkage algorithm is the maximum distance (or a minimum of the similarity) calculated between any two objects in the two clusters. It is the opposite of the single-linkage method.

- **Average-linkage:** Offers the intermediate metric, in-between the single-linkage metric and the complete-linkage metric.
- **K-Means clustering:** This algorithm uses a partitioning method as opposed to hierarchical methods. It dismembers the space crossed by the observations utilizing seeds chosen before starting the clustering process.

11.1.3 *Introduction*

Clustering is often utilized in market segmentation and credit scoring. In market segmentation, buyers are split into different segments so a blend of selling strategies that maximizes company revenue could be formulated. Examples include credit scoring to identify fraud, or simply, a grouping of objects (countries, firms, individuals, organizations, etc.) that are like each other using pertinent and quantifiable attributes.

A cluster is a set of objects that resemble each other but are different from other objects in a different cluster. Therefore, clustering involves organizing objects into meaningful groups. Predetermined attributes or characteristics are used to organize the objects into clusters.

Clustering seeks to group objects or members into different clusters to identify groups that exhibit a significant extent of internal (within a cluster) similarity and external (between clusters) dissimilarity.

Clustering might be done for various reasons.

1. To distinguish objects with common variables for decision-making.
2. To process data for further analysis, recognizing outliers, and suggesting hypotheses of underlying relationships.
3. To identify the most representative record of the cluster.

11.1.4 *Fundamental Clustering Methods*

When doing clustering, we first define a proximity measure (either of similarity or dissimilarity) between the two members. The proximity measure is used to identify clusters of similar objects. Next, the variables of each cluster are profiled to delineate their differences in relevant dimensions.

11.1.4.1 *Hierarchical (Agglomerative) Clustering*

There are two general classifications of clustering algorithms.

In hierarchical (agglomerative) clustering, each member is at first treated as a singleton cluster. The nearest pair of singleton/clusters are converged to form a bigger cluster. Next, recalculate the similarities between the clusters, which helps to further group the objects. Clusters created in the earlier stages are nested in those created afterward. The process continues until the desired criterion is fulfilled.

In sum, hierarchical algorithms work as follows:

1. Calculate the proximity among the data objects. Repeat: Merge data objects closest to each other.
2. Update the proximity between the new and the original clusters until all objects are grouped to become one final cluster.

All hierarchical agglomerative algorithms are variants of this generic approach. The difference lies in how the linkage metrics (measures of similarity between any two clusters) are calculated and determine the merging of objects and clusters. The exact metrics will be discussed later. The divisive or top-down approach is a variant of the hierarchical procedure. Under that approach, the iterative cycle starts with one cluster, where all objects are grouped. In the following steps, dissimilar objects are then separated into smaller clusters.

11.1.4.2 *Partitioning Clustering*

In partitioning clustering, the number of clusters is fixed before performing clustering. Every predetermined cluster has a seed. The seed is the initial cluster center, that is, the centroid or prototype. The seed, being the initial cluster center, will attract similar objects, within a prescribed similarity as members in the cluster. After all the objects get allocated, the cluster centroids are refreshed. This allows the objects to be reallocated and clusters to be reorganized in subsequent iterations. The hierarchical construction process outlined in the previous section is not affected by such methods since it separates the object space while searching for the groupings.

A generic partitioning algorithm is summarized as follows:

1. Choose K seed points as initial centroids or prototypes.

2. Form K clusters by allocating each object to its closest centroid.
3. Refresh the centroid of each cluster until the stopping criterion is fulfilled.

11.1.4.3 *Hierarchical vs. Partitioning Clustering*

Figure 11.1 shows pictorially the difference between hierarchical clustering and partitioning clustering.

The hierarchical clustering algorithm starts with all objects being grouped as single clusters or singletons. The objects are subsequently grouped based on their distances, in an agglomerative manner.

For the partitioning method, the number of clusters is fixed at the start and the clustering process allows members to move as long as the number of clusters is unchanged.

The same final solutions may be obtained by both the hierarchical and partitioning processes. However, it is notable that the clustering processes and concepts are fundamentally different.

Clustering analysis is an exploratory procedure. Although the clustering rules may be apparent, the clustering solution relies on the ability of the analyst to tag the clusters meaningfully. Different solutions may be achieved using different stopping rules which determine when the process ends. In many cases, more than one solution will be appraised before the clustering solution is determined.

Hierarchical Clustering   **Partitioning Clustering**

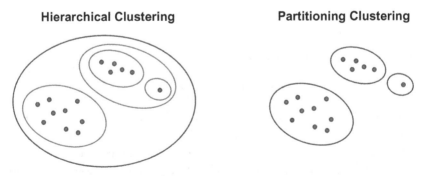

Figure 11.1:   Hierarchical clustering vs. partitioning clustering.

11.1.5 *Proximity Measure*

11.1.5.1 *Euclidean Distance*

The Euclidean distance (or metric) between two objects, x_i, and x_k, is defined as:

$$d_e(x_i, x_k) = \sqrt{(x_i - x_k)^T (x_i - x_k)}$$

$$= \sqrt{\sum_{j=1}^{p} (x_{ij} - x_{kj})^2}$$

$$\times d_S^2(x_i, \bar{x}) \leq (x_i - \bar{x})^T S^{-1}(x_i - \bar{x})$$

where p is the dimension of the space or the number of clustering variables.

The Euclidean distance is a non-negative scalar. If the objects are close to (far apart from) each other, then d_e is small (large). It is also symmetric, i.e., $d_e(x_i, x_k) = d_e(x_k, x_i)$ and satisfies the triangular inequality: $d_e(x_i, x_k) \leq d_e(x_i, x_m) + d_e(x_m, x_k)$ for all $x_i, x_i, x_i \in X$.

When the distance is evaluated with reference to the mean vector (centroid) of the cluster, the resultant quantity is called the error sum of squares:

$$\bar{x} = [\bar{x}_1, \bar{x}_2, \ldots, \bar{x}_p]^T$$

where $\bar{x} = [\bar{x}_1, \bar{x}_2, \ldots, \bar{x}_p]^T$ is the mean vector constructed from n objects and p clustering variables. This equation is used for evaluating within-cluster variations. In general, a cluster is expected to have a low sum of square error.

11.1.5.2 *Mahalanobis Distance*

The Mahalanobis distance is defined as:

$$d_S^2(x_i, \bar{x}) = (x_i - \bar{x})^T S^{-1}(x_i - \bar{x})$$

where $S = [s_{ij}]$ is an $n \times n$ symmetric sample covariance matrix that describes the variations and interrelationships of the clustering variables. The elements s_{ii}^2 and s_{ij}^2 refer to the variance and covariance between x_i and x_j, respectively. Effectively, this relation is used to standardize the observations as t-values. Being scale-invariant, this statistical measure should be used when it is important to capture the correlation between the variables to determine the distance between observations.

11.1.5.3 *Minkowski Distance*

The Minkowski distance is another common metric, where:

$$d_M(x_i, x_k) = \sqrt[M]{|x_i - x_k|^T |x_i - x_k|}$$

where M is a natural number and $||$ is the absolute value.

This distance measure may be regarded as a general form for quantifying distance between objects with the Euclidean distance being the case where $M = 2$. By varying M, we vary the weights given to small vs. larger differences.

11.1.5.4 *Determining the Number of Clusters*

The interpretation of cluster solutions builds on the required final number of clusters to a great extent, although there may not be an objective or standard selection procedure that can justify the number.

A straightforward method to decide the number of clusters is to delve into the clustering quality of the different solutions. For instance, if there is a huge increment in the average within-cluster variations (or error sum of squares) at step t, the previous solution at step $t - 1$ may be picked to be the final solution. Although it is not onerous, it is a practice that is useful.

A related set of standards gives a more accurate way to decide the final clustering solution. Known as the root-mean-square standard deviation (RMSSTD), the statistic pools the variations of all the clustering factors forming the cluster at each stage of the iterative process.

$$\text{RMSSTD}^2 = \frac{\text{Pooled sum of squares for all clustering variables}}{\text{Pooled degrees of freedom for all the clustering variables}}$$

A homogenous (or heterogeneous) cluster is characterized by a small (large) RMSSTD.

Notwithstanding, there is no generally acknowledged measure to determine what is "small" or "large". Nonetheless, the use of RMSSTD makes the analysis more tractable and may help in choosing whether a "final" solution is attained.

Different rules like Akaike's Information Criterion (AIC) and the Bayesian Information Criterion (BIC) may likewise be applied. However, a potential limitation is that these calculations may be more complicated. This is because the two standards require the user to know the underlying probability distribution of the observation. In any case, some analysts find such model selection criteria valuable when it is important to decide on the final number of clusters.

11.1.6 *Algorithms for Hierarchical Clustering*

There are various algorithms possible for hierarchical clustering. Different algorithms may bring about different solutions, but results should show a similar solution as long as some natural groupings are underlying the data. More than one clustering algorithm can be used to conduct an exploratory sensitivity analysis.

11.1.6.1 *Ward's Linkage*

In hierarchical clustering analysis, Ward's linkage can be utilized to minimize the total within-cluster sum of squares. The clusters are incrementally formed and the solution that results will feature the smallest sum of squares (or termed as error sum of squares (ESS)) in clustering.

For a given cluster k, denote ESS_k as the error sum of squares of all objects in the cluster from the cluster mean (i.e., centroid). If there are currently m such clusters, the total ESS is $\sum_{k=1}^{m} ESS_k$. At every step of the clustering process, we consider the union of every possible cluster and choose the one with the smallest ESS. Given the way the proximity coefficient is computed using Ward's algorithm, it is also known as the incremental sum of squares method. Further, Ward's method initially regards all objects as unique clusters, which makes the individual ESS zero.

11.1.6.2   *Centroid Linkage*

Centroid clustering is more formally defined using the squared Euclidean distance. For a cluster of m objects, a centroid is defined as:

$$\text{Centroid} = \left(\frac{\sum\limits_{i=1}^{m} x_{i1}}{m}, \frac{\sum\limits_{i=1}^{m} x_{i2}}{m}, \ldots, \frac{\sum\limits_{i=1}^{m} x_{ip}}{m} \right)$$

The centroid is a vector of mean values of the objects in that specific cluster. The centroid cluster approach forms clusters based on the shortest centroid distance between them. Effectively, the centroid distance is calculated as the distance between two clusters. More specifically, $d_E^2(\text{Cluster}_m, \text{Cluster}_k) = d_E{}^2(\text{Centroid}_m, \text{Centroid}_k)$ where $m \neq k$. A new centroid is obtained when an object or a cluster is grouped.

11.1.6.3   *Single-Linkage*

In the single-linkage algorithm, the proximity (or similarity) of the two clusters is represented by the minimum distance between any two objects in the clusters under all possible combinations. The smallest distance between any two objects of the clusters (one from each cluster) is computed.

When clusters m and k are selected for merging $d_E^2(\text{Cluster}_m, \text{Cluster}_k) = \min[d_E^2(\text{Object}_m, \text{Object}_k)]$, where Object_m is an object in cluster m and Object_k is an object in cluster k.

11.1.6.4   *Complete-Linkage*

The complete-linkage algorithm is the opposite of the single-linkage method. The proximity of two clusters is defined as the maximum of the distance (or minimum of the similarity) between any two objects in the two clusters, under all possible combinations.

In essence, $d_E^2(\text{Cluster}_m, \text{Cluster}_k) = \max[d_E^2(\text{Object}_m, \text{Object}_k)]$ where Object_m is an object in cluster m and Object_k is an object in cluster k.

The algorithm merges Cluster m with Cluster k if the distance between them is the smallest. Effectively, the Complete-linkage method endeavors to isolate clusters by maximizing the separation distance between them.

11.1.6.5 *Average-Linkage*

The average-linkage method (or group-average hierarchical clustering) gives the intermediate method, between the single-linkage and the complete-linkage methods. This method uses the average pairwise distances between all possible pairs of objects in each cluster as the proximity coefficient. A new cluster is formed by merging two clusters with the lowest average distances.

There are two types of group-average methods. The first is based on the between-group evaluation of distances. $d_E^2(\text{Cluster}_m, \text{Cluster}_k) = \text{average}[d_E^2(\text{Object}_m, \text{Object}_k)]$. Where Object_m is an object in cluster m and Object_k is an object in cluster k. Cluster m is combined with cluster k if the distance between them is the smallest.

On the other hand, the within-cluster algorithm considers: $d_E{}^2(\text{Cluster}_m) = \text{average}[d_E{}^2(\text{Object}_i, \text{Object}_j)]$ where Object_i and Object_j are within-cluster m and $i \neq j$.

Comparing this method to the between cluster algorithm, the within-cluster method is concerned more with separating clusters through minimizing intra-cluster distances.

11.1.7 *Partitioning*

11.1.7.1 *K-Means Clustering*

K-means clustering algorithm uses a partitioning method. It is different from the hierarchical methods, as it separates the space extended by the observations using seeds selected before starting the clustering process.

11.1.7.1.1 Initialization of Seeds

One crucial observation is that the clustering process is reliant on the initial assignment rule. If the initial rule assigns numbers 1, 3, 5 to Cluster 1 and the rest to Cluster 2, the clustering process would end in only one iteration. This suggests that the critical step lies in the choosing of the initial seeds. Most software initializes the clustering randomly or selects well-separated seeds.

To avoid getting a bad initialization, the software may do multiple runs, with each run using a different set of random centroids. Another method can also be adopted, which is to begin by taking the centroid of the whole sample. Next, for each successive initial seed, select the seed that is the farthest away from any previously defined initial seeds. This ensures that the centroids are randomly chosen and well-separated. This process aims to avoid convergence to local minima.

Bibliography

King, R. S. (2015). *Cluster Analysis and Data Mining: An Introduction.* Chaps. 1–4. Mercury Learning, Duxbury, MA.

Rousseeuw, P. J. and Kaufmann, L. (2005). *Finding Groups in Data: An Introduction to Cluster Analysis.* Chap. 1. Wiley, New Jersey, NJ.

Gan, G., Ma, C., and Wu, J. (2007). *Data Clustering: Theory, Algorithms, and Applications, SIAM.* Chaps. 1, 6, 7, and 9. American Statistical Association, Alexandria, VA.

Johnson, R. A. and Wichern, D. W. (2002). *Applied Multivariate Statistical Analysis.* Pearson, JASA, VA.

Ward, J. H. (1963). Hierarchical grouping to optimize and objective function. *J. Am. Stat. Assoc.*, 58, 236–244.

11.2   Sample Questions

Please select the most appropriate response.

Question 1

Movie Recommendation systems are an example of:

(a) Clustering only
(b) Clustering and Classification
(c) Clustering, Classification, and Learning

Question 2

What is the minimum no. of variables/ features required to perform clustering?

(a) 0
(b) 1
(c) 2

Question 3

For two runs of K-Mean clustering will give the same clustering results?

(a) Yes
(b) No
(c) Sometimes

Question 4

Is it possible that the assignment of observations to clusters does not change between successive iterations in K-means?

(a) Yes
(b) No
(c) Can't say

Question 5

Which of the following can act as possible termination conditions in K-Means?

1. For a fixed number of iterations.
2. Assignment of observations to clusters does not change between iterations. Except for cases with a bad local minimum.
3. Centroids do not change between successive iterations.

(a) 1 and 2
(b) 1 and 3
(c) 1, 2, and 3

Question 6

Which of the following clustering algorithms suffers from the problem of convergence at local optima?

(a) K-Means clustering algorithm
(b) Agglomerative clustering algorithm
(c) Both (a) and (b)

Solutions

Question 1

Solution: Option **c** is correct.

Generally, movie recommendation systems cluster the users in a finite number of similar groups based on their previous activities and profile. Then, at a fundamental level, people in the same cluster are given similar recommendations. In some scenarios, this can also be approached as a classification problem for assigning the most appropriate movie class to the user of a specific group of users. Also, a movie recommendation system can be viewed as a reinforcement learning problem where it learns by its previous recommendations and improves future recommendations.

Question 2

Solution: Option **b** is correct.

At least a single variable is required to perform clustering analysis. Clustering analysis with a single variable can be visualized with the help of a histogram.

Question 3

Solution: Option **c** is correct.

K-Means clustering algorithm converges on local minima that might also correspond to the global minima in some cases but not always.

Question 4

Solution: Option **a** is correct.

When the K-Means algorithm has reached the local or global minima, it will not alter the assignment of data points to clusters for two successive iterations.

Question 5

Solution: Option **c** is correct.

All three conditions can be used as a possible termination condition in K-Means clustering:

1. This condition limits the runtime of the clustering algorithm, but in some cases, the quality of the clustering will be poor because of an insufficient number of iterations.

2. Except for cases with a bad local minimum, this produces a good clustering, but runtimes may be unacceptably long.
3. This also ensures that the algorithm has converged at the minima.

Question 6

Solution: Option **a** is correct.

Out of the options given, only the K-Means clustering algorithm has the drawback of converging at local minima.

PART IV
Financial Innovations

Chapter 12

The Fourth Industrial Revolution

The technological revolution has been changing the way we live, work, and connect, at the scale, scope, and complexity that is unlike anything humankind has ever experienced before. Briefly, the First Industrial Revolution leveraged water and steam power to mechanize production. The Second harnessed electric power for mass production. The Third used electronics and information technology to automate production. Building on the Third, the Fourth Industrial Revolution is characterized by a fusion of advanced technologies that disrupt almost every industry in every economy (Guan *et al.*, 2020). The term "Fourth Industrial Revolution" is coined by Klaus Schwab of the World Economic Forum (WEF). It is an era of human–machine symbiosis where human capabilities are enhanced by machines, and machines acquire human-like characteristics. The WEF used the term "cyber-physical systems" to describe this phenomenon (Moyland, 2020). The Fourth Industrial Revolution, like its predecessors, will result in unemployment (widely forecasted as a temporary one) but will eventually bring about a higher quality of life as humans begin to think about human values.

12.1 Learning Objectives

- Explain what the Fourth Industrial Revolution is and its impact on businesses.
- Appraise characteristics of organizations and workforce to thrive in Industry 4.0 era.

12.2 Main Takeaways

12.2.1 *Main Points*

- The Fourth Industrial Revolution is an era where the fusion of technologies revolutionize how business solve problems, enhance human capabilities, and accelerate social and financial inclusion.
- Businesses should rethink their overall business strategy to be agile, resilient, address climate change, and include plans to upskill or retrain their workforce.
- The level of technological advancements has led to immense challenges for legislators and regulators.

12.2.2 *Main Terms*

- **Fourth Industrial Revolution:** A phenomenon that began at the start of the 21st century and was built on the digital revolution.
- **Fusion of technologies:** Various emerging technologies such as artificial intelligence, big data, cloud computing, blockchain, and IoT complement each other and create new growth opportunities.
- **Four durable shifts:** A definition by the World Economic Forum, defining shifts that businesses should be ready to adapt to—agility and customer-centricity, supply chain resilience, speed and productivity across the workforce, and eco-efficiency.

12.3 Introduction

The Fourth Industrial Revolution (also widely known as "Industry 4.0" or "4IR") helps enhance the global social welfare as well as the quality of life of people around the world. It brings about many impacts to individuals, businesses, as well as policymakers. The fusion of technologies enhances human capability, promotes inclusivity, and contributes to financial and social inclusion. Technologies, such as artificial intelligence, cloud computing, 5G, and mesh network, connect the disconnected, enable more people to be brought into the ecosystem, allowing businesses to better and more efficiently serve the un-served and underserved. It changes consumer behavior and the way people view ownership, data and digital footprint, and privacy for individuals. Consequently, this poses a challenge for policymakers to review and craft policies on using emerging technologies.

According to the World Economic Forum (2018), the Fourth Industrial Revolution focuses on the following areas:

- Autonomous and Urban Mobility.
- Artificial Intelligence and Machine Learning.
- Blockchain and Distributed Ledger Technology (DLT).
- Digital Trade and Cross-Border Data Flows.
- Drones and Tomorrow's Airspace.
- Precision Medicine.
- Internet-of-Things and Connected Devices.

On the other hand, Pricewaterhouse Coopers (PwC) identified eight essential technologies that are deemed as those that mattered most for businesses. They are: artificial intelligence, augmented reality, blockchain, drones, Internet-of-Things, robotics, virtual reality, and 3D printing (PWC Global, 2017).

While the past three Industrial Revolutions are essentially a prolongation of its predecessor, the Fourth Industrial Revolution has a more distinctive identity of its own. This is due to three reasons as follows. First, the speed of current breakthroughs has no historical precedent. Compared with previous industrial revolutions, the Fourth is evolving exponentially rather than at a linear pace. Second, it is disrupting almost every industry in every country. Finally, the breadth and depth of these changes herald the transformation of entire systems of production, management, and governance (Schwab, 2015).

The Fourth Industrial Revolution brings about "a fusion of technologies that is blurring the lines between physical, digital, and biological spheres" (Schwab, 2015). For instance, Smart City's concept is a fusion of various information and communication technologies, IoT devices, automation, data analytics, and artificial intelligence. The fusion of technologies will enhance our quality of living, create new markets, and present new growth opportunities in innovation.

The World Economic Forum's Global Lighthouse Network (McKinsey & Company, 2020) comprises leading companies that have successfully adapted to Industry 4.0 at scale, at individual sites, or across end-to-end supply chains. The Global Lighthouse Network

(GLN) aims to become a platform to share and learn from best practices in the search to improve sustainability, competitiveness, and customer satisfaction. The 54 companies (at the time of writing) in the GLN are from a wide range of manufacturing contexts across different industries, including oil and gas, consumer goods, automotive, medical devices, electrical components, home appliances, industrial equipment, and other industries. These companies draw competitive advantages from innovating their production systems. In September 2020, Alibaba, Micron Technology, Midea Group, Unilever, Groupe Renault, Janssen, Johnson & Johnson, Novo Nordisk, Saudi Aramco, DCP Midstream, and Schneider Electric joined the GLN.

One of the companies in the GLN is Tata Steel. Tata Steel deployed predictive maintenance on critical plant machinery. The predictive maintenance system is deployed on 25 types of equipment, and it complements the company's own internal maintenance strategy. The system can accelerate ramp-up, reduce cost, and average downtime with a mostly low-experience workforce (Boer, 2019).

Another case study presented in the GLN report is Unilever (Hefei). Unilever's top five use cases include:

- Digital-enabled automatic material call-off system.
- End-to-end real-time supply chain visibility platform.
- Supplier material delivery by e-Kanban.
- Lights-off packing.
- Artificial intelligence-enabled safety management.

Among these use cases, the implementation of a real-time end-to-end supply chain visibility platform has reduced order-to-delivery lead time by 50%, and its lights-off packing increases overall equipment effectiveness by 30%. Also, artificial intelligence-enabled safety management has reduced unsafe behavior by 80% (World Economic Forum, 2018).

12.4 Challenges and Opportunities

Following in the footsteps of previous revolutions, the Fourth Industrial Revolution can raise global income levels and enhance the standard of living. To this day, the Fourth Industrial Revolution has made significant advancements in serving the underserved, especially those with no prior access to digital services. Technology has also brought about a wider array of products and services that enhance the convenience and the quality of our lives. These services may include simple, mundane tasks such as the booking of public transport, accessing entertainment devices, or even making payments.

In time to come, technology-driven innovation will enable a supply-side miracle with advances in our economy's production capacity. This advancement is characterized by a lower cost of production for sectors such as trade and logistics. The advancement in technology will also create new growth sectors that are vital for a new era of the digitalized economy.

In 2020, the COVID-19 pandemic brought about uncertainties in demand and disruptions in the global supply chain. In line with the ongoing call for greener finance to address climate change and the call for environmental-friendly technologies, there is a need for a more inclusive, sustainable, and resilient world. Moreover, there is also a cultural shift toward a more open and sharing economy to lower entry barriers across many industries (Moyland, 2020). This calls for a "Great Reset" across sectors. According to WEF, the Great Reset needs companies to prioritize the needs of a broader set of stakeholders, thereby creating value for the society at large and, at the same time, contribute to ecological sustainability (Betti and Boer, 2020). Consequently, this will require companies and policymakers to balance encouraging innovations while ensuring proper accountability to the broader set of stakeholders.

12.4.1 *Impact on Individuals*

Without a doubt, the Fourth Industrial Revolution changes not only what we do but also who we are. It affects our perceptions of identity, privacy, ownership, human relationship, career development, and health. For instance, since 2019, Cisco's Consumer Privacy Survey (Cisco, 2020) has identified a group of "Privacy Actives" who care about privacy, are willing to act to protect it, and have already acted by switching providers over their data privacy policies. The emergence of Bitcoin blockchain as a decentralized peer-to-peer payment system revolutionizes our payment systems. However, it also causes us to raise questions such as self-sovereign banking (Johnson, 2019), trusted third party as a security loophole (Szabo, 2001), and trade automation that would potentially make global supply chains more effective (Deloitte, 2020). The increasing use of IoT devices in Smart City or smart home systems, while bringing convenience and comfort, invites questions on cybersecurity, and user privacy concerns. Data digitalization improves productivity and efficiency in serving consumer needs via automation and analytics. However, with digitalization, data can be easily replicated and shared without the authorization of its owner. Big data analytics produce precise results about an individual and intrude upon one's privacy. Data analytics performed on third-party cloud servers further aggravates the situation.

For investors, the Fourth Industrial Revolution presents both opportunities and risks. The breadth and depth of its impact and the resulting systemic changes require one to rethink the relevance of traditional methods in identifying specific areas of exponential innovation and disruption. Existing market indices are also found to not precisely capture the areas of innovation. The average tenure of a company in the S&P 500 was 15 years in 2015, while it was 65 years in 1920. Following the projection, forty percent of the current Fortune 500 companies will not be around in 2025.

To this end, perhaps a market- or outcome-based classification scheme is a more suited approach (Moyland, 2020). For instance, the total addressable market (TAM), serviceable available market (SAM), and serviceable obtainable market (SOM) models can be considered (Graham, 2020). Kensho Technologies, LLC, developed a classification framework called the "S&P Kensho New Economy Indices" (S&P Dow Jones Indices, 2020) to capture the industries and innovation of the Fourth Industrial Revolution. Ultimately, the importance is to consider the entire ecosystem, including the companies and the stakeholders they serve.

12.4.2 *Impact on Businesses*

The introduction of new technologies enables businesses to reach out to previously unreachable market segments. It also creates new and innovative ways to serve customer needs and allows businesses to leverage global digital platforms for research and development, marketing, the removal of intermediaries, enhancing production efficiency, and providing customized products or services to customers.

From the consumer perspective, with growing information transparency and digital consumer engagement, consumers are now demanding more customized touchpoints, virtual or physical, forcing businesses to adapt the way they design, market, and deliver goods and services. For example, the "sharing" or "on-demand" economy combines both demand and supply to disrupt existing industry structures. Commercial vehicle producer SAIC Maxus links their online order configuration tool to the production plant, suppliers, and the sales-and-distribution network. This innovation allows customers to indicate their needs, get the information propagated immediately to all relevant parties, and obtain their vehicles faster. As a result, the company grew 25% faster than the market (Boer, 2019).

A key trend in the business world is the availability of technology-enabled platforms. These platforms disintermediate demand and supply in conventional industries. Examples of these disruptive platforms include "sharing" or "on-demand" economies. The reason behind the development of this trend is credited to the development of smartphones and big data. With easier access to consumers, the future of consumer goods and services provided will be distinctively differentiated.

Also, technology lowers the cost of operation for business entities. This lowers the entry barriers to many business sectors, hence liberalizing the entry for many more platform businesses. This transformation seeks to change the business world significantly by introducing many more providers of new products and services.

In a recent Deloitte report (Deloitte Insights, 2020), companies that thrive are those that understand technologies, correctly apply them in the relevant business areas, including overall business strategy, workforce and talent strategies, societal impact, and technological operations. This allows the companies to innovate and grow and to attract and train the workforce needed.

The WEF report (Betti and Boer, 2020) listed *four durable shifts* occurring in the manufacturing and supply chain sector. They are:

1. *Agility and customer-centricity*: This requires businesses to quickly recognize changes in customer preferences and, thereafter, perform adaptive adjustments to manufacturing and production flows. The main driver of this change is demand uncertainty, which leads consumers to have greater access to information due to digitization and consequently demand better, more personalized products and services.
2. *Supply chain resilience*: In light of recent disruptions in global manufacturing and supply chain due to the pandemic, a highly resilient supply chain requires "connected, reconfigurable n-tier supply ecosystems, regionalization and an overall higher level of customization" (Betti and Boer, 2020).
3. *Speed and productivity across the workforce*: Businesses should recognize the need for lifelong learning and the importance of constantly reskilling the workforce while incorporating higher levels of automation. Also, businesses should continuously review workforce mobility and the ability to work remotely.
4. *Eco-efficiency*: The increased concern on climate change and the environmental impact of human activities necessitates businesses to adopt eco-friendly practices to respond promptly to new regulatory requirements.

The transformation from the Third Industrial Revolution to the current Fourth Industrial Revolution stresses the importance of digitalization. Indeed, companies could not rely on simple technologies anymore and must rely on a sufficiently innovative combination of technologies to enhance their business operations. In today's economy, the only constant is change, and business leaders are forced to continuously re-invent their previously successful businesses. Many business leaders had since risen to this challenge by reexamining their existing business models to navigate the current digitalized world's intricacies.

12.4.3 *Social Responsibility and Workforce Development*

In the same report, Deloitte reported evidence that businesses are beginning to find a balance between profit and purpose. Almost six in ten CXOs said that making a positive societal impact is among their top five desired outcomes of their Fourth Industrial Revolution investments. This includes reducing carbon emissions and consumptions, encouraging the reuse of materials, and eliminating wastes. Many businesses have also viewed climate change as one of their top priorities since they believe that climate change may hurt business operations. The focus on making a positive societal impact includes better opportunities to generate revenue and address pressure from external stakeholders (investors and customers) and employees.

For example, the United States and China are deploying drones to replace traditional inspection of oil and gas pipelines and high-voltage cables. A study by Deloitte (Deloitte, 2021) identified and quantified the technologies that will help businesses and governments to achieve the United Nations' Sustainable Development Goals (SDGs). According to the study, when deployed correctly, technologies could accelerate progress toward the development goals by 22%.

Following closely behind the desire to make a positive societal impact, businesses have started to recognize the importance of ensuring continual training and upskilling of their workforce. According to the survey, more than 80% of CXOs have created or are creating a corporate culture of lifelong learning and cited that executives not understanding the need to have updated skills necessary to thrive in the Fourth Industrial Revolution is a challenge to their businesses. However, there are a few problems that businesses face in this aspect. Firstly, many do not understand whether the senior executives or the IT staff should be trained in the technologies. The second problem is to identify the skills that will be needed in the future.

12.4.4 *Impact on Government*

Governments can use emerging technologies to improve the delivery of their services. For instance, blockchain's immutable and transparent characteristics can potentially revolutionize the provision of secure digital identity, healthcare recording, voting, and other government services. As mentioned previously, drones can be used to replace traditionally manual inspections at dangerous and hard-to-reach sites such as oil and gas pipelines. However, it is crucial to consider the technology risks before adoption.

As the digital and nondigital worlds continue to integrate, new technological platforms will enhance active citizenry in societies.

This includes more interaction between civil societies, more channels for citizens to provide feedback, and public authorities' monitoring. However, digital power can be a double-edged sword. There are ethical concerns when such technology is deployed in control of society or even when it comes to surveillance technology. Therefore, governments need to propose a more nuanced and acceptable approach when it comes to deploying technologies. Also, with the advancement of technologies and the decentralization of information, it is vital for governments to continuously recalibrate their approach when it comes to policy-making and engaging with the masses.

Previously, the Second Industrial Revolution had provided policymakers with the luxury of time to calibrate and implement policies in a top-down framework. This top-down and linear approach to decision-making had influenced our current systems of policy-making. However, this decision-making model is losing its relevance today. This is due to the speed of change, scope, and the profound impact of the Fourth Industrial Revolution on our existing systems. For instance, 3D printing may allow consumers to print drugs at home, leading the government to consider new regulatory issues (Schwab *et al.*, 2018).

The development of new technical standards, the formation of professional bodies to oversee technology development and adoption, crafting policies on organizations' relationships with competitors and stakeholders, ensuring that organizations are responsible for the environmental impact are among the responsibilities that need to be undertaken by the government. Formulating regulations in these areas requires the government to work closely with the industry.

In July 2019, France introduced a new digital tax which applies a 3% levy on revenues from digital services earned in France by companies with more than €25 million ($28 million) in French revenue and €750 million ($838 million) worldwide (Peccarelli, 2020).

The proposals to tax digital goods and services based on the locations where they are delivered are necessary from an economic standpoint. However, this autonomy will imply a very confusing set of policies resulting from each jurisdiction imposing different taxation policies. This mess can be a source of potential conflict.

Undoubtedly, the level of technological advancements has led to immense challenges for legislators and regulators. One of the key challenges lies in the fair balance between regulation and technological innovation and the simultaneous creation of enough capacity to encourage innovative development. The solution to this is to adopt a form of governance model known as "agile governance". This agility model is not new. It had been developed by private sectors for the field of software development and business operations. However, agility comes with responsibilities. Regulators have to cooperate closely with business and civil sectors to upgrade their regulations in this time of rapid changes continuously.

12.5   Industry 4.0 and the Workforce

Industry 4.0 has brought us various benefits. However, it does have a fair share of disadvantages. The most severe disadvantage is social inequality. As we rely more on machines and less on human labor, the existing gap between return on capital and return on labor may increase. In the era of Industry 4.0, the most vital component for the industry will not be capital or labor. Instead, it will be innovation and the ability to generate new ideas. Talents are most likely to replace capital as the most critical factor of production in the future. Indeed, people with ideas will be the most sought-after resources. This will lead to a market that is increasingly divided. Low-skill and low-wage menial tasks will be replaced by machines, while the higher-paid jobs requiring innovation are less likely to be replaced. This enhanced level of dichotomization could potentially trigger a greater level of social tensions.

Such a scenario is labeled by Keynes (1930) as "technological unemployment". Many studies have observed that future employment growth in both developed and developing countries follows a U-shape curve. Employment opportunities are high for low- and high-skilled workers but low and declining for middle-skilled workers. Examples of this category are pervasive in routine cognitive and manual skills, such as factory and clerical workers (Autor *et al.*, 2006; Goos and Manning, 2007; Autor, 2015b).

All Industrial Revolutions have led to economic transformation, where productivity is largely improved and employment is threatened. The productivity improvement has led to an improved standard of living overall, although it takes time to materialize

(Chuah *et al.*, 2018). The period when employment falls is dubbed "Engels' pause" (Allen, 2009).

The Fourth Industrial Revolution is slightly different from all previous Industrial Revolutions. While the previous Industrial Revolutions used technology to improve production, mass-produce, and automate productions, the Fourth Industrial Revolution leverages technology and machines to perform non-routine tasks that had been hitherto reserved for humans. The applications of machine learning, neural networks, and deep learning to perform big data analysis and discover hidden patterns in consumers' data and the deployment of chatbots to replace customer service officers have sparked unease in the labor force.

However, it is worth noting a few points. First, machines' replacement of labor cannot happen instantaneously and is highly dependent on the specific contexts. The cost needed to put these technologies in place may outweigh the labor costs in certain developing countries. Second and more importantly, Autor (2015a, 2015b) stressed the strong relationship between humans and machines in complementing each other as machines and algorithms, though efficient, excel only in conducting tasks in specific domains.

To this end, note that many organizations agree that they are looking toward hiring for "mindset" rather than skills. Open and flexible mindsets with a desire to learn will thrive in this ever-changing era. All in all, successful implementation of Industry 4.0 technologies will benefit the workforce in the long-term.

12.5.1 *Assessing Impact of Technological Innovation on Jobs and Wages*

Acemoglu and Autor (2011) and Acemoglu and Restrepo (2018) developed a framework that allows us to assess how technological innovations will impact our jobs and wages. In their publications, the two types of innovations, namely "enabling technologies" and "replacing technologies", benefit humans in the sense that productivity is expanded. This leads to higher employment and wages and also spurs new complementary tasks. Educational institutions and policymakers should identify the complementary skills that are future-proof to mitigate the negative effects of technological change. To achieve future-proof skills, policies that encourage mobility and agility are highly recommended. Also, there must be a sufficient safety net and channels to upskill workers to acquire new skills needed for the new economic sectors.

Policies that increase the cost of labor unnecessarily will exacerbate the problem of the adoption of technologies that displace labor. Hence, labor market reform should emphasize improving workers' skills and mobility across different sectors or geography. A desirable business environment is necessary for the sustainability of the labor market. This helps to reduce market failures. However, these policies should be developed with the motive of enhancing efficiency. They should not be implemented with a protectionist mindset or to preserve outdated sectors.

Apart from improving mobility in the labor market, sufficient protections are necessary as well. These protections include safety nets, such as cash rebates, to support the poor and the unemployed. Also, financial assistance to these workers' families should also be considered. There is currently evidence from the World Bank, which suggests that well-designed safety nets effectively reduce poverty and inequality and protect workers against creative destructions.

Ultimately, in the long run, distribution policies that include a broader target audience should be considered. These include investing in more inclusive public goods, social insurance, and perhaps, a universal minimum wage scheme. The idea of inclusion is to allow society to share the collective benefits of a digitalized society instead of confining the benefits to a minority (Freeman, 2015).

12.6 Future Benefits and Challenges

Experience from previous industrial revolutions indicate that to realize the full benefits of the new technologies, the world must meet three pressing challenges (Schwab *et al.*, 2018):

1. *Ensure fair distribution of the benefits of the Fourth Industrial Revolution.* The reasons that a subset of the world population may miss out on the benefits include service unavailability, inability to afford, or subtle system biases that tend to privatize profits.
2. *Manage externalities in terms of risk and harm.* The vulnerable populations, natural environment, and future generations are

often overlooked and left unprotected from suffering due to unintended consequences. For instance, under some scenarios, quantum computing could create risks to privacy and security to the adopters of cryptographic schemes; autonomous vehicles could increase road congestion in an already-crowded city.

3. *Ensure that the Fourth Industrial Revolution is human-led and human-centered.* To be human-centered is to empower humans as a meaningful agency in the world. New technologies in Industry 4.0 can analyze and make decisions based on data that humans cannot process and, at some point, in ways that humans cannot control or understand. Via digital networks, the effect and impact will spread far more quickly than any previous technological development phase.

Klaus has highlighted four key principles useful in defining a mindset that is adaptable to 21st century challenges.

1. *Think systems, not technologies.* A good system design that focuses on delivering well-being is more important than a "good technology". Without a good system design, technology can do more harm than good.

2. *Empower, not determine.* Human decision-making should be valued. Technologies should be harnessed in a way that gives us more choices and more control over our own lives instead of a human being influenced by the decisions computed by the machines.

3. *Think by design, not by default.* Design thinking and systems thinking approaches will help us to understand the structures that guide the world and how technologies may shift systems into new configurations.

4. *Think of values as a feature, not a bug.* Technologies are not just tools; they have their implicit values from how and why they are developed.

Bibliography

Acemoglu, D. and Autor, D. H. (2011). Skills, tasks and technologies: Implications for employment and earnings. In *Handbook of Labor Economics*, Volume 4, Eds. Ashenfelter, O. and Card, D. E. Amsterdam: Elsevier.

Acemoglu, D. and Restrepo, P. (2017). Robots and jobs: Evidence from US labor markets. NBER Working Paper 23285, National Bureau of Economic Research, Cambridge, MA.

Acemoglu, D. and Restrepo, P. (2018). The race between man and machine: Implications of technology for growth, factor shares, and employment. *Am. Econ. Rev.*, 108(6), 1488–1542.

Allen, R. C. (2009). Engels' pause: Technical change, capital accumulation, and inequality in the British industrial revolution. *Explor. Econ. Hist.*, 46(4), 418–435.

Autor, D. H. (2015a). Polanyi's paradox and the shape of employment growth. In *Re-Evaluating Labor Market Dynamics. Proceedings-Economic Policy Symposium-Jackson Hole*, 2014. Federal Reserve Bank of Kansas City.

Autor, D. H. (2015b). Why are there still so many jobs? The history and future of workplace automation. *J. Econ. Perspect.*, 29(3), 3–30.

Autor, D. H., Katz, L. F., and Kearney, M. S. (2006). The polarization of the U.S. labor market. *Am. Econ. Rev.*, 96(2), 189–194.

Betti, F. and Boer, E. (2020). Global lighthouse network: Four durable shifts for a great reset in manufacturing. World Economic Forum White Paper, September 2020. Retrieved from http://www3.weforum.org/do cs/WEF_GLN_2020_Four_Durable_Shifts_In\Manufacturing.pdf.

Boer, E. (2019). *Lighting the Way: New Leaders in the Fourth Industrial Revolution*. McKinsey & Company. Retrieved from https://www. mckinsey.com/business-functions/operations/our-insights/operations-blog/lighting-the-way-new-leaders-in-the-fourth-industrial-revolution.

Chuah, L., Loayza, N., and Schmillen, A. (2018). The Future of Work: Race with- not against- the Machine. World Bank Group, Research & Policy Briefs From the World Bank Malaysia Hub, No. 16.

Cisco (2020). Protecting data privacy to maintain digital trust. Consumer Privacy Survey. Retrieved from https://www.cisco.com/c/dam/en_us/about/doing_business/trust-center/docs/cybersecurity-series-2020-cps.pdf?CCID=cc000742{&}DTID=esootr000515{&}OID=rptsc023 525.

Deloitte (2020). How blockchain can reshape trade finance. Retrieved from https://www2.deloitte.com/content/dam/Deloitte/global/Documents /grid/trade-finance-placemat.pdf.

Deloitte (2021). Digital with purpose: Delivering a SMARTer2030: Deloitte and GeSI launch a new report on the impact of digital technology on the SDGs. Retrieved from https://www2.deloitte.com/uk/en/pages/ strategy/articles/digital-with-purpose-delivering-a-smarter-2030.html.

Deloitte Insights (2020). The Fourth Industrial Revolution: At the intersection of readiness and responsibility. Retrieved from https://www2. deloitte.com/content/dam/Deloitte/de/Documents/human-capital/ Deloitte_Review_26_Fourth_Industrial_Revolution.pdf.

Freeman, R. B. (2015). Who owns the robots owns the world. *IZA World Labor*, 2015, 5. doi:10.15185/izawol.5.

Goos, M. and Manning, A. (2007). Lousy and lovely jobs: The rising polarization of work in Britain. *Rev. Econ. Stat.*, 89(1), 118–133.

Graham, A. (2020). TAM methodology: An explanation and example of total addressable market analysis. Retrieved from https://www.toptal .com/finance/market-sizing/total-addressable-market-example.

Guan, C., Jiang, Z., and Ding, D. (2020). *The Emerging Business Models*. World Scientific.

Johnson, A. (2019). *Self-Sovereign Banking Putting You Back in Control of your Money*. Forbes. Retrieved from https://www.forbes.com/sites/ alastairjohnson/2019/10/03/self-sovereign-banking-putting-you-back-in-control-of-your-money/?sh=2e44a4751f61.

Keynes, J. M. (2010). Economic possibilities for our grandchildren. In *Essays in Persuasion* (pp. 321–332). Palgrave Macmillan, London.

McKinsey & Company (2020). Global lighthouse network: Insights from the forefront of the 4th industrial revolution. Retrieved from https: //solutions.mckinsey.com/msd/global-iot/sites/default/files/CXO{%} 20summary{%}20-{%}20Global{%}20Lighthouse{%}20Network{%} 20-{%}20Jan{%}202020.pdf.

Moyland, J. (2020). *The Fourth Industrial Revolution: Are We Ready?* S&P Global. Retrieved from https://www.spglobal.com/en/ research-insights/featured/the-fourth-industrial-revolution-are-we-ready.

Peccarelli, B. (2020). *Bend, Don't Break: How to Thrive in the Fourth Industrial Revolution*. World Economic Forum. Retrieved from https: //www.weforum.org/agenda/2020/01/the-fourth-industrial-revolution -is-changing-all-the-rules/.

PWC Global. (2017). The essential eight. Retrieved from https://www.pw c.com/gx/en/issues/technology/essential-eight-technologies.html.

Schwab, K. (2015). The Fourth Industrial Revolution: What it means and how to respond. Retrieved from https://www.foreignaffairs.com/artic les/2015-12-12/fourth-industrial-revolution.

Schwab, K., Davis, N., and Nadella, S. (2018). *Shaping the Future of the Fourth Industrial Revolution*. The Crown Publishing Group.

S&P Dow Jones Indices. (2020). Equity: New economies. Retrieved from https://www.spglobal.com/spdji/en/index-family/equity/kensho-new -economies/{#}overview.

Szabo, N. (2001). *Trusted Third Parties are Security Holes*. Satoshi Nakamoto Institute. Retrieved from https://nakamotoinstitute.org/trusted-third-parties/.

World Economic Forum (2018). Centre for the Fourth Industrial Revolution. Retrieved from https://www.weforum.org/centre-for-the-fourth-industrialrevolution/areas-of-focus.

12.7 Sample Questions

Please select the most appropriate response.

Question 1

Which of the following statements is correct?

(a) Low-skilled workers will find their jobs increasingly replaced by machines.
(b) The employment situation will become less favorable for middle-skilled workers.
(c) High-skilled workers will find themselves in a tough job market due to lifelong learning and competition.

Question 2

The Fourth Industrial Revolution is different from all previous Industrial Revolutions in that:

— the pace of discoveries is significantly faster
— the more developed a country is, the more widespread is the disruption caused by it
— entire systems, rather than individual devices, of production, need to be replaced
— machines are increasingly performing capably in tasks categorized as cognitive

How many of the above statements are correct?

(a) 1
(b) 2
(c) 3

Question 3

In the four durable shifts identified by WEF, the agility and customer-centricity mean:

(a) To react quickly in recognizing the impact of consumers' activities on climate change
(b) To react quickly to changes in consumer preferences and perform adaptive adjustments to own manufacturing and production flows
(c) To identify the changes in consumer preferences quickly using data analytics and artificial intelligence algorithms

Question 4

Which of the following is not a problem faced by businesses in workforce development?

(a) Identify how to train the staff
(b) Identify who to train in the technologies
(c) Identify the skills that will be needed

Question 5

How should the governments and regulators react to changes brought about by the Fourth Industrial Revolution?

(a) Use a top-down framework to calibrate and implement policies
(b) Craft policies with a balance between and technological innovation
(c) Create a one-for-all policy across all industries

Solutions

Question 1

Solution: Option **b** is correct.

Many studies have observed that future employment growth in both developed and developing countries follows a U-shape curve, where employment opportunities are high for low- and high-skilled workers, but low and declining for middle-skilled workers.

Question 2

Solution: Option **c** is correct.

The second point is incorrect since it is disrupting almost every industry in every country. Disruption of less developed countries can also be widespread as shown by the example of M-Pesa.

Question 3

Solution: Option **b** is correct.

Agility and customer-centricity require businesses to quickly recognize changes in customer preferences and thereafter perform adaptive adjustments to manufacturing and production flows. This change's main driver is demand uncertainty, which leads consumers to have greater access to information due to digitization and, consequently, demand better, more personalized products and services.

Question 4

Solution: Option **a** is correct.

Many corporations have created a corporate culture of lifelong learning. The problems they face in this aspect are that many do not understand whether the senior executives or the IT staff should be trained in the technologies and the way to identify the skills that will be needed in the future.

Question 5

Solution: Option **b** is correct.

The top-down framework is losing its relevance due to the velocity, scope, impact of Industry 4.0 in our existing systems. The advancement of technologies and decentralization of information require governments to continuously recalibrate their approach when it comes to policymaking and engaging with the masses.

Chapter 13

Fintech and Financial Inclusion

Technological innovations are transforming the way financial services are provided. This phenomenon is commonly known as "Fintech". Some researchers have termed this wave of transformation as the "Fourth Industrial Revolution". Regardless of the terminology, fintech does have a profound impact on our society. This introductory chapter first introduces various advantages of fintech, such as LASIC and financial inclusion. Following this, specific case studies on the application of fintech by companies will be discussed to offer a more comprehensive view of fintech applications.

13.1 Learning Objectives

- Discuss how fintech can achieve financial inclusion.
- Appraise the characteristics of successful fintech businesses based on the LASIC framework.
- Criticize how fintech applications use technologies to improve productivity and efficiency of financial services.

13.2 Main Takeaways

13.2.1 *Main Points*

- Fintech is short for financial technology, and it refers to innovative financial services or products that are enabled by technology.

- Financial inclusion refers to the ability to provide financial services to previously underserved segments. In practice, financial inclusion is most prominent in the servicing of micro, small, and medium enterprises (MSMEs) and underserved individuals.
- Successful fintech businesses should exhibit the five important attributes highlighted by the LASIC principles. The LASIC principle is a succinct summary for a series of comparative advantages possessed by fintech firms over traditional financial institutions. It stands for low margin, asset-light, scalable, innovative, and compliance easy.

13.2.2 *Main Terms*

- **Fintech:** Innovative financial services or products that are enabled by technology.
- **Financial Inclusion:** A principle that refers to providing financial services to the previously underserved social segments.
- **LASIC Principles:** Five important attributes of business models that can successfully utilize financial technology to create a sustainable business for financial inclusion. They are low margin, asset-light, scalable, innovative, and compliance-easy.

13.3 Introduction to Fintech

Fintech is short for financial technology, and it refers to innovative financial services or products that are enabled by technology. In simple terms, it refers to the technology currently in place, which improves traditional financial services. This improvement can fall under three main simplified categories (1) convenience, (2) scope of services offered, and (3) the number of people to whom the services can now be offered. For illustrative purposes, the following examples are provided as follows:

1. *Convenience*: Mobile payment eliminates the need to carry a physical wallet. Also, mobile payment services can be utilized to make payments for household bills as a one-stop station. This enhances the level of convenience of daily users and could never be achieved with conventional technologies.

2. Traditionally, financial institutions provide a limited scope of services to consumers. With the introduction of fintech, not only can the platform now provide payment services to the consumers, but also a wide range of related services covering insurance, clinical, shopping, investment, and microfinancing. This enhances the scope of services provided by a single platform.
3. Financial institutions do not serve everyone due to the potentially high cost incurred. However, fintech services carry a much lower cost and can be provided to those underserved by financial institutions. These services include microloans to small business owners or mobile banking to remotely located farmers. This concept of extending services to more people is known as financial inclusion, which will be discussed later.

The exact definition of fintech is yet unavailable. But the general trend is that it results in an improvement in our daily lives. With the development of technology, an increasing number of people are now spending more time on mobile and the Internet, causing changes to consumer expectations and a higher volume of data. Thus, this drives the demand for up-to-date and innovative financial services and products. For example, the higher volume of data now collected can make a more accurate prediction in service offerings.

The reason why we switched to fintech providers is due to the repeated failure of traditional financial institutions. Back in 2008, members of the public were disappointed when taxpayers' money was used to bail out irresponsible financial institutions. The financial system relies heavily on the trust of the people and investors, which stood jeopardized after the global financial crisis. Technology is trusted and welcomed by consumers and entrepreneurs, and such alternative or innovative financial services can help rebuild public trust. This is the reason for the initial growth behind fintech.

In terms of application, the proliferation of e-commerce changes consumers' behavior, and many e-commerce platforms, such as Amazon and Alibaba are providing banking-like services to the consumers, which in turn expedite the trend of incorporating technologies to innovate financial services. Fintech can contribute to financial inclusion as well, providing useful and affordable financial services and products to unbanked or unserved individuals, households, and businesses that the traditional financial system is unable to do so. Both of these points will be elaborated under the subsequent sections.

13.4　Financial Inclusion

Financial inclusion refers to the ability to provide financial services to previously underserved segments. It is a key enabler to reducing poverty and boosting prosperity. The World Bank has identified several strategies to achieve financial inclusion (The World Bank, 2018). The strategies include modernizing retail payment systems and government payments and leveraging technology for financial inclusion. This essentially refers to the usage and application of fintech.

In practice, financial inclusion is most prominent in servicing micro, small, and medium enterprises (MSMEs). Currently, there are an estimated 200 million MSMEs in the emerging economies with no access to services offered by the formal financial system. As such, financial inclusion has emerged to be a critical driver of economic development. The methods of providing financial inclusion to MSMEs can be through providing payment solutions, investments, and credit services. In credit services, for instance, microfinancing was traditionally unavailable due to its lack of profitability. However, due to the lower cost sustained by fintech operations, microfinancing can be provided at a much lower cost and hence be profitable.

Another important segment that was underserved was individuals. Currently, an estimated 2 billion individuals are underserved by traditional formal financial institutions. Fintech provides inclusive services to this segment either through cheaper services, new services previously unavailable, or by providing value-adding services. A prominent example is the growth of mobile payment solutions. In developing nations, financial inclusion through mobile services had caught up with the developing world. This is due to the lack of physical financial institutions in these regions. Therefore, mobile banking and payment availability overcomes the limitation of traditional financial institutions and provides financial services to a population much in need of them.

In summary, how fintech provides inclusive services to underserved individuals and MSMEs can be summarized by the following three key points:

1. *Reducing the cost of services*: Fintech reduces the cost of providing services tremendously. Existing services can now be provided at much lower prices or simply for free. In sectors with high price sensitivity, this lower price will lead to a more than proportionate increase in demand. As a result, fintech companies have a comparative advantage in customer acquisition.

2. *Serving the underserved*: More often than not, there are gaps in the supply of services. Some segments of the consumers remain underserved due to the high cost of supply. Fintech, with its lower cost of production, can serve these underserved segments by providing new services to them.

3. *Offering solutions to improve services*: Through technology, fintech can provide better solutions that enable established business entities to serve their current customers better. This includes the use of big data and analytics to enhance their understanding of customers.

Following the LASIC principle (to be introduced later), a few key principles applicable in this section are the asset-light, low margin, and innovative nature of fintech. Through its innovative characteristics, it can provide disruptive services that differentiate from those offered by traditional banks. The low margin and asset-light nature also allow fintech to provide services with a much lower cost while serving a much larger group of users.

Overall, fintech had had tremendous progress in achieving global financial inclusion. According to Ernst and Young (EY)'s Fintech Adoption Index 2017, the average percentage of digitally active consumers using fintech services reached 33% across the 20 mature and developing markets. Across the 20 markets, China stood out significantly. Chinese consumers have the highest adoption rate of 69%.

In essence, fintech harnesses a wave of disruption, which changes the provision of products and services in the financial sector. This has forced many traditional financial institutions to change their current business model to compete with fintech. The key reason behind the successful and disruptive nature of fintech lies in its indifferent service model. Fintech is meant to serve the masses, not just privileged segments. This empowers them to reach out to a much larger service segment and attract more demand in the process. For example, fintech companies such as Ant Financial are capable of providing micro-loans to previously underserved SMEs. Through data analytics, Ant Financial can access the credit rating of borrowers through their payment records. Ant Financial can calibrate the number of loans that extend to these SMEs. Conveniently, this decision can be made within a day, easing SMEs' cash flow needs.

We will now take a closer look at the application of fintech in real life and how companies can actualize these realities in their business models.

13.5 Applications of Fintech

Consumers can benefit from fintech due to, but not limited to, the following four aspects: (1) reducing the prices of services, (2) providing a new type of value-adding service, (3) solving a problem for another business, and (4) focus on previously underserved customer segments (Chong *et al.*, 2020). An increasing number of startups are entering the realm of fintech. Payments, Robo-advisory, quantitative trading, crowdfunding, and peer-to-peer (P2P) financial services, insurance, blockchain, and regulation technology are among the most popular areas. We will take a look at each of these domains.

13.5.1 *Mobile Payment*

Mobile payments are third-party platforms that facilitate the payment or transfer of money through mobile phones. These platforms may include popular applications such as PayPal in the United States, WeChat Pay in China, and Paylah in Singapore. These services greatly transformed the way we make payments, not just to vendors but also for our utilities.

13.5.2 *Crowdfunding and Peer-to-Peer (P2P) Financial Services*

Raising capital has always been a vital issue for businesses. All business needs capital but not any business can raise capital easily. This is especially true for newly established SMEs without any loan history. Instead, crowdfunding allows small businesses to raise capital efficiently and cost-effectively. These platforms function as the intermediary between the borrower and the lenders. They facilitate the matching of desired borrowers with the respective lenders while charging a fee for this service. Some examples of P2P lenders are EasiCredit in China, MooLahSense in Singapore, and Lending Club in the United States. Traditionally, the provision of crowdfunding services carries high risks. However, fintech providers utilize algorithms to determine borrowers' payment history and default risks. This greatly reduces the default risks associated with MSMEs. Also, fintech platforms are asset-light, which confers to them lower costs of operation. This reduces the cost of providing micro-loans to SMEs as well.

13.5.3 *Quantitative Trading and Robo-Advisory*

Quantitative trading and Robo-advisors refer to the use of technology and big data to conduct trading. Robo-advisors are platforms for conducting trade, selecting assets, and managing the portfolio for the clients. Quantitative trading, on the other hand, refers to the use of historical data and algorithms to make technology investment decisions. More often than not, these two services are integrated on the same platform. Through the integration of both Robo-advisory and quantitative trading, these platforms are capable of offering sound investment advice and portfolio management services to individual investors. Compared to traditional portfolio managers, Robo-advisors charge a lower fee. The lower cost is extremely attractive to individual investors who are likely not endowed with a large sum of capital, nor can they afford professional wealth managers. Also, the digitalized platform is deemed attractive for millennials. Instead of consulting an advisor, investments can be made through a digitalized platform familiar to millennials. Apart from individuals, small business entities may also rely on Robo-advisors for asset management services. Similar to their counterparts, the cost of engaging in such services is much lower than engaging traditional banks. Some examples of Robo-advisors are Betterment and Wealthfront in the United States. These Robo-advisors allow an initial investment account with US$25 to start investing, unlike traditional investment advisors that need at least US$100,000, if not more.

13.5.4 *RegTech*

A critical component of the financial world is compliance. Regulatory technology, more commonly known as "RegTech" for short, utilizes technology to assist in the process of monitoring, reporting, and compliance (Ascent, 2020). There are two key components of RegTech that are vital in the execution of tasks. The first lies with big data analytics. Using data analytics, financial institutions can detect shortcomings in their current operations and adjust to comply with regulatory requirements. Another process lies with automation. Auditing or monitoring is labor-intensive. By automating the monitoring and compliance processes, risk identification and regulatory compliance could be made more efficient and accurate. Commercial banks and financial regulators have already taken the first step in adopting RegTech. In time to come, RegTech could be applied to a much wider base.

13.5.5 *InsurTech*

InsurTech refers to the line of technology that is utilized in the insurance industry. These technologies aggregate to improve the overall efficiency of the entire insurance industry. InsurTech has a few traits in attaining greater efficiency. First, through data analytics, InsurTech can allow insurance underwriters to gain more in-depth insight into the client's financial activities. This deeper understanding of the client can allow more customized solutions that are tailored for each client. Machine learning allows InsurTech to be highly innovative in meeting clients' needs. Second, besides customized products, InsurTech offers a more precise evaluation in terms of financial risks. By accessing the client's financial data, the default risks of the client could be further understood and managed. This benefits the insurance provider by lowering the risks associated with underwriting. The combined advantages of InsurTech have brought great challenges for traditional insurance companies. For example, the Chinese InsurTech firm Zhong An had gained extremely lucrative margins for its partner firm, Ping An, which is China's largest insurance provider. The key about Zhong An is that it operates solely online and can generate high premium by accumulating small margins. One year since its inception in 2013, Zhong An has assisted its partner Alibaba to increase sales by RMB 100 million and their freight insurance policies by 50%.

13.5.6 *WealthTech*

WealthTech stands for wealth and technology. It is an area that is gaining great traction over the years in assisting investors to manage their wealth (Cheng, 2019). WealthTech refers to digital solutions that aim to enhance the wealth management and investing process. To counter disruptions from tech-savvy startups and to cope with changes in consumer expectations, WealthTech uses technologies such as big data and artificial intelligence to help wealth management companies to tap into untapped markets. A prominent example of WealthTech includes Robo-advisor, which aims to provide wealth-related management advice using fintech. As one of the newer sectors of fintech, WealthTech has seen a recent downward trend in investment activity by about 62.5% from the year 2018 to 2019. This slowdown is the result of investors' re-focus on late-stage companies and safe bets. KPMG had estimated that this trend would continue into 2020 due to the impact of COVID-19 (Pollari and Ruddenklau, 2020).

13.5.7 *Digital Banking*

Digital banking simply refers to banking business models with digital or online services without any physical branches. There are many similar terminologies used for digital banking, such as virtual banking, online banking, mobile banking, or even social banking in Europe. A key point to note is that digital banking would not have been possible without smartphone proliferation. Smartphones offer users the convenience of tapping into their online digital banking account on the move. This has extended the reach and widened the target audience of digital banks. Currently, some notable digital banks have been started by fintech companies. More examples will be featured in the subsequent case study on Fidor. Fintech giants, such as Alibaba Group in China or Grab in Singapore, are expanding their digital banks rapidly. Both of these companies carry common characteristics. They provide a payment platform, a digital wallet, and a higher adoption rate. These key characteristics allow both entities to offer banking services quickly.

On the other spectrum, traditional banks are capable of dispensing digital banking functions too. Traditional banks are taking a closer look at promoting synergies, accelerating innovative opportunities, and providing more digital banking services. The new technology also underlines the need for regulators to keep up with an ever-changing banking landscape. Currently, China leads the digital banking sector and is one of the most financially inclusive countries globally. Some digital banks are being set up in Kenya, the United States, and Europe. But the sustainable model for growth rests with China because it has a social agenda behind digital banking. For digital banking customers, the top three global banks in 2013 were Chinese banks, namely, Industrial and Commercial Bank of China (over 100 million mobile banking customers), China Construction Bank (117 million), and Agricultural Bank China (83.0 million).

13.5.8 *Other Fintech Applications*

Besides the various domains of fintech discussed above, applications of fintech remain extensive. Fintech is also applied in areas such as lending, remittance, chatbots, and personal finance. Also, the Monetary Authority of Singapore (MAS) is embarking on a Central Bank Digital Currencies (CBDC) project (FinTech Singapore, 2020). This important application of fintech shows that fintech utilization is not limited to the business world but also the governmental authorities. Another key application of fintech is remittance. Remittance is classified as a form of cross-border payment. Cross-border payment technology is essential for facilitating trade. This is all the more important in an open economy that is widely found today, where cross-border trade is vital for economic growth. More information can be found in PWC (2019).

13.6 The LASIC Framework

Successful fintech businesses should exhibit the five important attributes highlighted by the LASIC principles as follows (Lee and Teo, 2015). The LASIC principles shown in Figure 13.1 constitute a

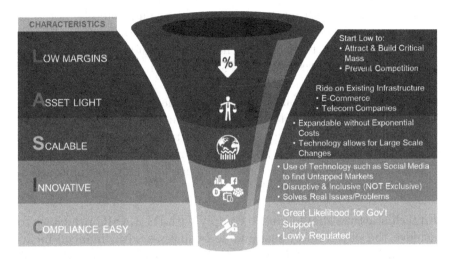

Figure 13.1: The LASIC principles.

succinct summary of the comparative advantages possessed by fintech firms over traditional financial institutions:

- *Low-profit margin*: It is a crucial aspect of a successful business, especially in the Internet and technology space, as many users expect information to be provided without any charges, which is plausible due to the user base and data amount being quite large. Network effects play an important role in a fintech business's success—whether it can get a large user base and profits via the high demand and low margins.
- *Asset-light*: In a rapidly changing environment, asset-light companies have more flexibility and capability to incorporate advancements in technologies and adjust their services and products without high fixed costs on assets. This way, startups can maintain a competitive edge among peers.
- *Scalable*: To better harness the benefits of network effects, fintech startups should be scalable, which means the technology used should incorporate some new changes to deal with potential changes or advancements in the future. Thus, the technology used should be carefully designed beforehand, and consideration of scalability cannot be neglected.

- *Innovative*: In addition to being asset-light and scalable, both the products and operations of successful fintech startups need to be innovative to meet consumers' changing demands and make full use of innovations in technologies to enhance profitability and ensure a sustainable business model.
- *Compliance easy*: Compliance and regulation costs are also key considerations. For fintech startups, compliance costs in a highly regulated environment may largely disrupt businesses' profitability and jeopardize the cash flows. In a compliance-easy environment, companies can spend less on compliance and more on innovation, which can promote growth.

Fintech startups should aim to comply with the LASIC principles to achieve the objective of sustainable businesses with financial inclusion. The last point of LASIC principles, compliance easy, is an external factor. Entrepreneurs can devote themselves to overcoming the pain points in existing systems, actively raising awareness, highlighting the importance of services to the public, users, and even the government to improve the regulatory environment and gain government support. This section offers several case studies of fintech companies that had undergone the re-bundling process. These companies are often successful fintech companies that later expanded to offer financial services. In summary, successfully "rebundled" companies often carry the following characteristics:

1. Large existing network of consumers.
2. Technology is seamlessly integrated with its social network.
3. Provide innovative and trusted fintech services.

13.6.1   *Case Study: Telecom Fintech—M-PESA*

An important factor that M-PESA leveraged was the rapid expansion in mobile penetration. More often than not, developing countries face challenges such as the lack of internet infrastructure or even smart devices. Hence, it is important to leverage simple mobile technologies. M-PESA capitalized well on this by leveraging on Short Message Services (SMS) to transfer money. Launched in 2007 by Safaricom, M-PESA had driven financial inclusion by providing money transfer services to an underserved segment and facilitated local payment as well as international remittance services. The success of M-PESA is striking. As of 2014, M-PESA had acquired 122,000 registered merchants and 19.3 million registered customers. This is a relatively high number considering that only 5 million Kenyans have access to a bank account. Also, M-PESA had expanded to other regions as well as include other services. Geographically, M-PESA had expanded to countries such as Tanzania, Afghanistan, South Africa, India, and Eastern Europe.

13.6.2 *Case Study: Social Media Fintech—Fidor*

In contrast to developing countries, developed countries are blessed with much better internet connectivity. This has led to very different business models for the fintech companies located in the developed countries. One such example is Fidor Bank. Fidor Bank differentiated itself as the first bank that operated through the internet and on social media. Established in Germany back in 2007, Fidor Bank expanded to include 300,000 registered users and 250,000 community members in 2014. Beyond the number of users, Fidor Bank has achieved €200 million worth of deposits and extend €160 million worth of lending in total. What distinguishes the operation of Fidor from the conventional bank is its low overhead cost. Being on social media platforms, Fidor Bank need not maintain a large network of branches. Hence, its operation can be sustained with only 34 staff. Also, it only costs the bank €20 to set up a customer with full banking services. This cost is once again much lower than conventional banks. As a leader in innovative banking processes, Fidor Bank had attained various awards back in 2013 and was distinguished as the "most innovative bank for social media" in Germany.

Besides its low overhead, Fidor Bank is also successful in facilitating the establishment of a banking community, payment service, and solution provider software known as "fidorOS". The software contributed significantly in facilitating banking services on social media by allowing the transfer of money on social media, providing lending services, and crowd financing services. To serve business establishments, Fidor Bank had provided another application known as "Fidor TECS". The application is very similar to iTunes provided by Apple. Fidor TECS helps to facilitate a community for its business-to-business clients.

13.6.3   Case Study: E-commerce Fintech—Alibaba Group

Alibaba Group is a much larger business entity compared to both Fidor bank and M-PESA. The Alibaba Group is a Chinese fintech services provider founded by Jack Ma and his team back in 1999. Being a much larger services provider, Alibaba provides a range of consumer-to-consumer (C2C), business-to-consumer (B2C), and B2B sales services via the Internet. Alibaba started as a simple yet innovative payment platform known as "Alipay" in 2004. Alipay was an important payment device that facilitates trust in using a third-party custodian. The third-party custodian will safely keep the consumers' money and only credit the seller once the consumers had received the products. Such a service gained popularity quickly in China and had since expanded to capture a large consumer base. This large base of consumers allowed Alibaba group to expand its services to include other sectors such as utilities, credit, health care, and even insurance. Currently, Alibaba is capable of providing offline services through POS systems used by small business entities.

13.6.4   The CLASSIC Framework

CLASSIC was created by one of the four audit firms EY (see Figure 13.2). C denotes customer-centric with simple-to-use and high-convenience products and services with needs-focused propositions designed around particular consumer use cases and pain points. There is thus a high degree of customer engagement. L stands for being legacy-free with purpose-built systems designed around digital channels and fulfillment. There is little drag from discontinued products, before-acquisition, or regulatory liabilities. A stands for asset-light as in small fixed asset base to create significant operating leverage. As a result, balance sheets are frequently rented out or outsourced to other parties. S stands for scalability. Scalability is built into the business model by leveraging partnerships, distribution, and simplicity with low capital requirements. The next S also stands for a fundamentally simple customer proposition as highly focused and transparent in business processes. I stands for innovative, with innovation across the spectrum in the new business models, products, services, and delivery modes. Finally, the last C stands for

Common characteristics	Description
Customer-centric	▸ Simple, easy-to-use, high-convenience products and services ▸ "Needs-focused" propositions designed around particular consumer use cases and pain points ▸ High degree of customer engagement
Legacy-free	▸ Purpose-built systems designed around digital channels and fulfilment ▸ Little drag from discontinued products, prior acquisitions or regulatory liabilities
Asset light	▸ Low fixed-asset base creating significant operating leverage ▸ Balance sheet frequently rented or outsourced to other parties
Scalable	▸ Scalability built into the business model by leveraging partnerships, distribution and simplicity ▸ Low capital requirements
Simple	▸ Fundamentally simple customer proposition ▸ Highly focused and transparent business processes
Innovative	▸ Innovation across the spectrum, e.g., new business models, products and services and delivery models
Compliance light	▸ Simple and unbundled models that are often designed so as to avoid the need for authorisation

Source: Chuen and Teo, EY analysis

Figure 13.2: The CLASSIC framework.

compliance light as simple, unbundled models that are often designed to avoid the need for approval (Lee and Teo, 2015).

Bibliography

Ascent (2020). What is RegTech? Retrieved from https://www.ascentregt ech.com/what-is-regtech/.

Cheng, M. (2019). *The Future of Wealthtech*. Forbes. Retrieved from https://www.forbes.com/sites/margueritacheng/2019/02/19/the-future-of-wealthtech/?sh=6308e50235e6.

Chong, G., Jiang, Z., and Ding, D. (2020). *The Emerging Business Models*. World Scientific.

FinTech Singapore (2020). Singapore FinTech Report 2021. Retrieved from https://fintechnews.sg/wp-content/uploads/2021/01/Singapore-Fintech-Report-2021-Alibaba-Cloud-Fintech-News-SG-.pdf.

Lee, D. K. C. and Teo, E. G. S. (2015). Emergence of FinTech and the LASIC principles. *J. Financial Perspect.*, 3(3), 1–26.

Pollari, I. and Ruddenklau, A. (2020). *Pulse of FinTech H12020.* KPMG. Retrieved from https://assets.kpmg/content/dam/kpmg/xx/pdf/2020 /09/pulse-of-fintech-h1-2020.pdf.

PWC (2019). Crossing the lines: How fintech is propelling FS and TMT firms out of their lanes. Global Fintech Report 2019. Retrieved from https://www.pwc.com/gx/en/industries/financial-services/assets/pwc -global-fintech-report-2019.pdf.

The World Bank (2018). Financial Inclusion: Financial Inclusion is a key enabler to reducing poverty and boosting prosperity. Retrieved from https://www.worldbank.org/en/topic/financialinclusion/overview {#}2.

13.7 Sample Questions

Please select the most appropriate response.

Question 1

Which of the following principles are embedded within the LASIC principles?

(a) Low-profit margin, asset-heavy, scalable, innovative, compliance issue

(b) Low-profit margin, asset-light, scalable, innovative, compliance easy

(c) Large profit margin, asset-light, scalable, innovative, compliance easy

Question 2

Your client approached you with a query. He wanted to know which type of fintech services he should explore in the context of monitoring, regulatory, reporting, and compliance. Which of the following represents the best answer?

(a) RegTech
(b) Crowdfunding and P2P Financial Services
(c) InsurTech

Question 3

Which of the following options provide the best description of services provided by Fidor?

(a) A P2P Lending Platform specializing in micro-loans for small enterprises
(b) A social media FinTech, which is also the world's first online-only bank
(c) An insurance provider in China that specializes in using AI to predict risks

Question 4

Which answer best represents the combination of reasons as to how fintech provides inclusive services to underserved individuals and MSMEs?

 I. Offer a previously paid-for service free of charge or at a cheaper rate
 II. Provide a new type of value-adding service
III. Solve a problem for another business

(a) I only
(b) I and II
(c) All of the above

Question 5

Which of the following is true about RegTech?

(a) RegTech is the use of technology to automate regulatory processes
(b) RegTech is the use of technology to help businesses manage regulatory compliance
(c) RegTech is the use of technology to enhance efficiency in court proceedings

Solutions

Question 1

Solution: Option **b** is correct.

All positive principles are incorporated. Profit margin is low since many users expect information to be provided without any charges and the user base and data amount are quite large.

Question 2

Solution: Option **a** is correct.

Regulatory technology (RegTech) describes the use of technology in the context of regulatory monitoring, reporting, and compliance, particularly in the financial services industry.

Question 3

Solution: Option **b** is correct.

In social media fintech, Fidor Bank was established in Germany in 2007. It is the world's first online-only bank that operates through the Internet and using social media.

Question 4

Solution: Option **c** is correct.

Question 5

Solution: Option **b** is correct.

Chapter 14

Emerging Technologies

This introductory chapter first introduces the technologies used in fintech, which include big data, artificial intelligence, cloud computing, blockchain, and the Internet of Things (IoT). Also, the applications of each technology will be discussed.

14.1 Learning Objectives

- Discuss the development of various emerging technologies and each relates to fintech.
- Examine how businesses use emerging technologies to enhance production.

14.2 Main Takeaways

14.2.1 *Main Points*

- Artificial intelligence refers to developing and applying corresponding computer systems to perform tasks that usually require human intelligence to complete. These include visual perception, speech recognition, gesture control, machine learning, language processing, etc.
- In general, blockchain is a public database (or distributed ledger) based on blockchain technology. Blockchain refers to the combination of data exchange, processing, and storage technologies formed by multiple participants based on modern

cryptography, distributed consistency protocol, peer-to-peer network communication technology, and smart contract programming language.

- Cryptocurrencies refer to digital currencies where encryption techniques and consensus algorithms are used to determine the monetary policy in the absence of a custodian, such as the central bank. Bitcoin is the first and has been one of the most prominent cryptocurrencies used by traders in the crypto field.

- Cloud computing provides usable, convenient, and on-demand network access to users, allowing them to enter a configurable computing resource sharing pool with a charge-based model. Computing resources include networks, servers, storage, application software, and services. Cloud computing users can quickly acquire computing resources with little management effort and little interaction with service providers.

- Big data refers to large-scale data collection. Massive data scale, rapid data flow, diverse data types, and low-value density are the four salient characteristics of big data. The big data processing process can be divided into five steps: data acquisition, data storage and management, data processing, data analysis and visualization, and data display.

- The IoT refers to a network that comprises various sensor devices (according to Internet protocol) capable of gathering and sharing electronic information over the Internet, which can realize monitoring, connection, and interaction. It aims to recognize the connection between objects and objects as well as objects and people in a network. The IoT's industrial chain can be divided into four major layers: sensor layer, network layer, platform layer, and application layer.

14.2.2 *Main Terms*

- **Big Data:** A reference to large-scale data collection. Massive data scale, rapid data flow, diverse data types, and low-value density are the four most relevant big data characteristics.

- **Artificial Intelligence (AI):** The branch of computer science that emphasizes the development of intelligent machines that can think and work like humans.

- **Blockchain:** A public database (or distributed ledger) that enables peer-to-peer transactions without involving any third parties such as central banks or financial institutions (or decentralized).
- **Cloud computing:** A branch of computing that provides usable, convenient, and on-demand network access to users, allowing them to enter a configurable computing resource sharing pool with a charge-based model.
- **Internet of Things (IoT):** refers to a network that comprises various sensor devices (according to Internet protocol) capable of gathering and sharing electronic information over the Internet, which can realize monitoring, connection, and interaction.

14.3 Artificial Intelligence

Artificial intelligence refers to developing and applying corresponding computer systems to perform tasks that usually require human intelligence to complete. Artificial intelligence mainly focuses on these six areas: computer vision, image recognition, machine learning, natural language processing, speech recognition, and robots.

Artificial Intelligence (AI) is about making agents (i.e., machines or computers) mimic cognitive functions that would require the human mind or human intelligence. It has evolved from early symbolic AI to knowledge-based expert systems, machine learning, and today's deep learning. Artificial Intelligence is a general field, Machine Learning is a subset of AI, and Deep Learning is a branch of Machine Learning. In sum, AI has delivered significant effectiveness and progress over the last decade because of big data availability, dramatically improved advanced algorithms, and the powerful computing ability and cloud-based services.

Based on the logical architecture, AI can be divided into the following three layers:

- The *basic layer* provides a fundamental guarantee for the realization of artificial intelligence at the hardware and theoretical level. This mainly includes AI chips and deep learning algorithms.

- The *technical layer* is based on the basic layer's support and is designed to perform tasks that normally require human intelligence in the past.
- The *application layer* is based on the capabilities of the technical layer to solve specific real-life problems. For example, computer vision technology can be used in face recognition in the security field, and natural language technology can be used in intelligent customer service.

Machine learning algorithms can be categorized into four major types: supervised, unsupervised, semi-supervised, and reinforcement learning. Below is a summary of these four. More will be covered in the fintech module in Level 1.

The *supervised learning* algorithms are trained/taught using given data records. The data are labeled, which means that the desired output for input is known. For example, a credit card application can be labeled as either approved or rejected. The algorithm receives a set of inputs (the applicants' information) along with the corresponding outputs (whether the application was approved or not) to foster learning.

Unlike supervised learning, in *unsupervised learning*, the algorithm is not trained/taught on the "right answer". The algorithm tries to explore the given data and detect or mine the data's hidden patterns and relationships. In this case, there is no answer key.

Semi-supervised learning is similar to supervised learning, as it is often being used to address similar problems. The difference between semi-supervised learning and supervised learning is that in semi-supervised learning, a small amount of labeled data and a large amount of unlabeled data are provided at the same time. Semi-supervised learning will be utilized when the labeling process is too costly for a fully labeled training process. With the labeled data, a large amount of unlabeled data can be classified by semi-supervised learning algorithms. On top of that, a new model will further be trained using the new labeled data set.

Reinforcement learning aims to find out the action which leads to the maximum reward or drive to the optimal outcome. A set of allowed actions, rules, and potential end states are provided to the machine beforehand, and the job of the machine is to explore different actions and observe resulting reactions.

Compared to machine learning, deep learning automates the feature engineering of the input data (the process of learning the optimal features of the data to create the best outcome) and allows algorithms to automatically discover complex patterns and relationships from the input data. Deep learning is based on artificial neural networks (ANNs), which were inspired by information processing and distributed communication nodes in biological systems, similar to human brains.

14.3.1 *Applications of AI*

Intelligent voice is a field of computer science that enables computers to listen and speak in the same way that a human does. Speech recognition and voice synthesis are currently their core applications. The current development of intelligent voice technology has been relatively mature. Intelligent voice interaction is rapidly becoming a mainstream human–computer interaction mode.

Computer vision is a field of computer science that works on enabling computers to see, identify, and process images in the same way that human vision does and then provide appropriate output. Image restoration, enhancement, segmentation, identification, and manipulation are key computer vision tasks. It is like imparting human intelligence and instincts to a computer. With the breakthrough of computer vision technology in multiple fields in recent years, such as the release of Generative Adversarial Networks (GANs) in 2014 spawned multiple and diverse applications, it has become the most popular technical branch of artificial intelligence.

Natural language processing (NLP) is a field of computer science that works on enabling computers to understand and write languages. Natural language understanding and natural language generation are the main natural language processing tasks. Natural language processing is a key technology to realize cognitive intelligence. Although it still faces great challenges, its future progress and breakthrough will be of considerable significance to human society.

Artificial Intelligence uses computer science to solve problems in the same way that humans do. Typical application scenarios include AlphaGo, smart game player AlphaStar, anti-fraud and anti-money laundering in the financial field, Robo-advisory, and automatic trading.

14.4   Blockchain

In general, blockchain is a public database (or distributed ledger), and it enables peer-to-peer transactions without involving any third parties such as central banks or financial institutions (or decentralized). Blockchain refers to the combination of data exchange, processing, and storage technologies formed by multiple participants based on modern cryptography, distributed consistency protocol, peer-to-peer network communication technology, and other features such as smart contracts. These technologies are combined in a new way to realize the tamper-proof data, chain structure traceability, trusted point-to-point transmission, etc. The blockchain infrastructure aspect is mainly composed of six layers: data layer, network layer, consensus layer, incentive layer, contractive layer, and application layer, as shown in Figure 14.1.

The development of blockchain can be divided into three stages.

- *Stage 1*: Application to currencies and payments. Blockchain is the technology that underpins bitcoin, the first cryptocurrency that enabled a decentralized, peer-to-peer exchange of value (payments).

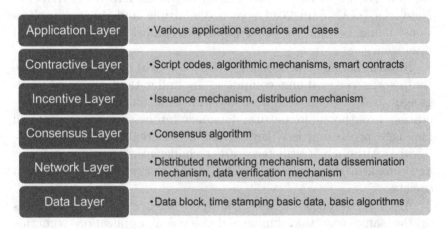

Figure 14.1:   The six blockchain layers.

- *Stage 2*: Utilization in the financial industry. Many financial services players have created cryptocurrency-based services using smart contracts.
- *Stage 3*: Expansion to areas such as social notarization and intellectualization.

14.4.1 *Applications of Blockchain*

Supply chain: Using blockchain technology, relevant commodities data across stakeholders can be tracked and recorded at any point in time. This is particularly crucial in the supply chain sector due to the multiple change-of-hands amongst stakeholders. The adoption of blockchain technology can also ensure the authenticity of food production and expiry dates, along with other data to ensure food safety.

Healthcare: Blockchain technology can enable secured information symmetry, reduce fraudulent medical supplies and improve the efficiency of healthcare administration. It can improve the healthcare system's efficiency as it can aggregate information of patients into a common database to facilitate secure and timely cross-institution data sharing even for the new patients. The problem of counterfeit vaccines and drugs can also be solved using blockchain technology. The authenticity of medicines can be verified by identifying a unique code recorded on the blockchain, which cannot be tampered with. In terms of healthcare administration, blockchain can improve efficiency by automating the execution of related procedures via smart contracts, thereby reducing administrative costs and saving time for both patients and healthcare providers. Afterward, the processed data can be used for insurance claims and bill management conveniently, reducing administrative costs, mitigating risks such as insurance fraud, and improving verification efficiency.

Public Service: Blockchain technology has various applications in the public service sector to improve government efficiency and security. For instance, blockchain can enable governments to disseminate trusted and official information that is identified with a digital signature. Milestones can be recorded and stamped, showing the progress of the project and the use of funds. This will improve transparency, allow citizens to verify and monitor government projects, and promote trust in the public sector.

Education: It is possible to adopt blockchain technology in multiple aspects of the education sector, such as record-keeping and certificate generation. Students' information and credentials can be stored and recorded digitally on a blockchain, so do the diplomas, certificates, and transcripts. This way, the information is securely recorded, challenging to alter, and can be easily verified.

Intellectual Property: Blockchain technology can record and protect authors' copyright and reduce copyright authentication and verification costs. Also, smart contracts can be used to allow authors to authorize others to use their work with certain fees paid and automatically track and distribute the income. At present, many blockchain copyright platforms provide digital assets and copyright trading functions, enhancing the efficiency of digital asset circulation. In the future, offline physical assets can also be registered and traded on the blockchain, allowing for more use cases.

Personal Identity Protection: With blockchain technology, personal data can be protected, and identity theft can be reduced. With personal data stored on a blockchain network, hackers would no longer have a single point of entry (if it is not a private blockchain), and they are disadvantaged by the requirement to alter hundreds of thousands of data stored in multiple agencies or nodes to hack an individual's data. Even if data was compromised, hackers would only have access to a small fraction of one's data due to the decentralized nature of blockchain, making the process of stealing identities exponentially challenging as opposed to accessing a centralized database.

14.5   Cryptocurrencies and Bitcoin

Cryptocurrencies refer to digital currencies where cryptographic techniques and consensus algorithms are used to determine the monetary policy in the absence of a custodian, such as the central bank. The blockchain is the underlying technology that provides the security of the cryptocurrency. There are close to 850 cryptocurrencies available, and some notable examples are bitcoin, ether, and DASH. Bitcoin is recognized as a commodity by the Commodity Futures Trading Commission in 2015, giving it a status just like oil and gold. Bitcoin is a digital currency that is not printed by a central bank, thereby making it unaccountable to any political powers or population. The value of bitcoin is derived from the supply and demand economics and is free from politicized monetary policy, that is, the devaluation of a country's currency to curtail inflation or pay for national debts.

Bitcoin first came into the scene with the publication of the paper "Bitcoin: A Peer-To-Peer Electronic Cash System" by Satoshi Nakamoto in November 2008. In January 2009, the Bitcoin network came into existence with the first open-source Bitcoin client's release and the issuance of the first bitcoin to Satoshi Nakamoto. Bitcoin is created digitally by a community of developers, users, investors, and miners. Anyone is welcome to join the Bitcoin network either as a developer, investor, user, or miner. There are no administrators in the Bitcoin network to vet a person's credibility. Bitcoin is, therefore, a permissionless blockchain network. Despite the absence of administrators to safeguard the Bitcoin network, the network is safe from hackers due to the use of cryptography and blockchain technology. As the Bitcoin network is designed to run without any administrators, all rules and protocols regarding how the network works are all coded into the network by developers. Bitcoin network participants must follow the rules to maintain a state of consistency across the network, thus guaranteeing the security of the network.

14.6   Cloud Computing

Cloud computing provides usable, convenient, and on-demand network access to users, allowing them to enter a configurable computing resource sharing pool with a charge-based model. Computing resources include networks, servers, storage, application software, and services. Cloud computing users can quickly acquire computing resources with little management effort and little interaction with service providers.

Cloud computing gathers scattered computing resources to form a shared resource pool, organized and deployed through the network to realize large-scale information processing and optimize information processing efficiency. In the traditional deployment framework, the enterprise's system framework deployment, operation, and maintenance consume a lot of cost and time.

However, cloud computing technology, computing, storage, and networks can be virtualized to form a database that can quickly implement product deployment, reduce management and construction processes, and improve corporate operating efficiency.

Cloud computing has the following characteristics:

1. *Strong computing ability*: The "cloud" has a considerable scale and many servers, which gives users supercomputing power.
2. *High reliability*: The "cloud" uses measures such as fault tolerance of multiple copies of data, homogeneous computing nodes, and interchangeability to ensure high reliability of services, making cloud computing more reliable than local computers.
3. *Pay on demand*: Cloud computing can build a huge resource pool through virtualization technology, in which its system can perform metering services and can detect, measure, and even control the use of resources. At the same time, users can use related resources according to their own needs and pay based on demand.
4. *Resource-sharing*: The virtualization technology of cloud computing maps a computer's physical resources such as servers, networks, memory, and storage into virtual resources, thereby installing and deploying multiple virtual machines to achieve multi-user physical resources sharing.
5. *Low-cost*: Cloud computing's automated centralized management frees a large number of enterprises from being burdened by the increasingly high data center management costs, and its versatility significantly improves the utilization rate of resources compared to earlier traditional systems. This enables users to enjoy the low-cost advantage of the "cloud" fully.

There are three types of cloud: public cloud, private cloud, and hybrid cloud that combine both. Cloud services can be categorized as IaaS (Infrastructure as a Service), PaaS (Platform as a Service), and SaaS (Software as a Service). Cloud computing enables a more advanced financial system by providing low-cost and high-efficiency infrastructure and is largely used in financial industries such as insurance and banking that utilize big data analytics and demand high computing power.

14.7 Big Data

Big data refers to large-scale data collection. Massive data scale, rapid data flow, diverse data types, and low-value density are the four salient characteristics of big data.

Under the background of the rapid development of mobile Internet and information explosion, mobile hard disk capacity has grown from being measured in gigabytes (GB) to terabytes (TB). The volume of data processed by Internet companies has jumped from petabytes (PB) to exabytes (EB), and even zettabytes (ZB). For example, the amount of data processed daily by Baidu, China's largest search engine, is nearly 100PB, equivalent to the information content of 5,000 national libraries.

Many traditional tools and methods are no longer suitable for processing such big data. Before this big data era, most data were stored in a structured way, so a relational database was frequently used. However, since the emergence of the concept of "big data", data are stored in semistructured and even unstructured forms. Traditional techniques requiring high cost are inefficient and are gradually unsuitable for this much larger volume of data.

The big data processing process can be divided into five steps: data acquisition, data storage and management, data processing, data analysis and visualization, and data display. Structured or semistructured raw data are first collected from browsing histories or transaction records, etc. Then the raw data are stored and managed using Hadoop Ecosystem or NoSQL. After that, users can access the processed data for their purposes, such as decision support, business intelligence, and recommendation systems, etc. Finally, the data can be visualized and displayed for presentation.

In recent years, various sensor devices, such as PC and mobile phones, have rapidly increased. Under the background of the Internet's popularization and the infiltration of the IoT, the rapid growth of data provides a massive database for the big data industry. The emergence of chips such as GPU, FPGA, and TPU provides a hardware basis for processing big data. The emergence of new technologies such as cloud computing, artificial intelligence also provides technical support for the development of the big data industry. Cloud computing can realize on-demand charging, reduce the difficulty and cost for an enterprise in applying big data, enable enterprises to build extraordinary big data solutions, and promote big data industries. Artificial intelligence improves the efficiency of big data algorithms through cutting-edge algorithms such as deep neural networks.

Data resources in the financial services sector are relatively obtainable, and the business development function is highly dependent on this data. Therefore, the application of big data technology in the financial services field starts early and develops rapidly. The application of big data in the financial industry has been widespread, more mature than other industries, and has achieved remarkable results. Data "assetization" is becoming increasingly prominent, and in-depth big data analysis is becoming more important. User portraits and knowledge graphs have become essential technologies.

14.7.1 *Application of Big Data*

Big data is widely used in the financial services industry. With asset quality assessment and business recommendation as to the core, it has developed a broad application landscape across the banking, securities, and insurance industries.

Big data technology is mainly used for credit risk assessment and supply chain finance in the banking sector. In terms of credit risk assessment, the traditional credit business is based on the customer's past credit data and transaction records, calculating the corresponding default probability, and deciding whether to approve the loan. By using big data analysis technology, banks can deeply integrate internal data (credit and transaction records) with external data (such as customer consumption information, credit records, etc.) to obtain multi-dimensional results. Banks can also obtain a conclusion closer to the facts with a combination of historical customer credit, industry development status, and real-time operational situation. In terms of supply chain finance, banks generate inter-enterprise relationship maps that indicate investing, controlling, lending, and guaranteeing relations between enterprises and the relationship between shareholders and legal persons.

In the securities industry, big data technology can be used for quantitative investment and Robo-advisory. In terms of quantitative investment, big data helps to make a more precise prediction. Securities companies can predict stock market prices and movement by analyzing daily stock market information and market sentiment. In terms of Robo-advisory, it can analyze the needs and risk profile of a customer and finally provide an investment recommendation. Simultaneously, with natural language processing (NLP) technology, Robo-advisors can answer standardized questions and reduce the work cost of financial customer service.

In the insurance industry, big data technology can be used for fraudulent insurance identification and risk pricing. In terms of fraudulent insurance identification, through the integration of internal and external data, the corresponding insurance fraud identification model can be established to monitor the abnormal data effectively, to pick out the suspected fraud claims that may be cases of "fraudulent insurance". In terms of risk pricing, taking vehicle insurance as an example, insurance companies can determine the current premium by collecting driving data, behavior data, and health data, etc., through intelligent monitoring devices installed on vehicles. At the time of claim settlement, picture assessment based on machine learning technology dramatically improves the efficiency of claims and reduces operating costs.

14.8 Internet of Things

The development of the Internet was represented first by the PC Internet (massive information was interconnected and shared through the Internet), followed by the mobile Internet stage, where people can communicate with each other. Now the third stage of Internet development is the IoT.

The IoT refers to a network that comprises various sensor devices (according to Internet protocol) capable of gathering and sharing electronic information over the Internet, which can realize monitoring, connection, and interaction. It aims to recognize the connection between objects and objects as well as objects and people in a network.

The IoT's industrial chain can be divided into four major layers: sensor layer, network layer, platform layer, and application layer. The sensor layer is mainly composed of chip modules and terminal equipment. The network layer and platform layer are mainly supported by communication and platform services. The application layer is mainly composed of data analysis and related applications led by Internet vendors, as well as vertical industry application services led by traditional vendors.

Communication technologies of the IoT include short-distance links, cellular links, and low-power wide-area links. Different linking technologies have different advantages and disadvantages, and the corresponding application scenarios are also different.

- *Wired communication technology*: Ethernet is the traditional technology for connecting devices in a wired local area network (LAN) or wide area network (WAN).
- *Wireless short-range communications technology*: Wi-Fi is a common in-home networking application. ZigBee is a low-cost, low-powered, and low-rate bidirectional communication technology.
- *Wireless long-range communications technology*: NB-IoT offers long-range, low power consumption, low delay sensitivity, and secure data transmission. LoRa is ideal for providing intermittent low data rate connectivity over significant distances.

The IoT can be applied in many aspects. The mobile Internet of Things refers to the extension and use of IoT in the mobile industry. It can be used for large-scale consumer applications such as intelligent hardware, smart home, car networking, and so on. The Industrial Internet of things (IIoT) refers to the extension and use of the IoT in industrial sectors such as industry, agriculture, energy, etc., which can be applied to building infrastructure in the industry and aid in industry transformation and upgrading.

The urban Internet of Things is about permitting cities to access real-time data-driven management that concerns urban systems, including water, waste, energy, and transportation. The deep learning algorithm, represented by the artificial neural network, is continuously popularized and optimized, allowing big data algorithms to efficiently process mass data and analyze the IoT's mass data. It enables the control and interaction to be more intelligent, while cloud computing technology can significantly reduce costs, thus promoting the development of the IoT.

Bibliography

Goodfellow, I., Bengio, Y., Courville, A., and Bengio, Y. (2016). *Deep Learning*. Cambridge: MIT Press, 1(2).

Guan, C., Jiang, Z., and Ding, D. (2020). *The Emerging Business Models* (Chapter 4: Introduction to Financial Technology). World Scientific, Singapore.

James, G., Witten, D., Hastie, T., and Tibshirani, R. (2013). *An Introduction to Statistical Learning*. New York: Springer, p. 112.

McAfee, A., Brynjolfsson, E., Davenport, T. H., Patil, D. J., and Barton, D. (2012). Big data: The management revolution. *Harvard Business Rev.*, 90(10), 60–68.

Ng, A. (2019). Machine learning yearning: Technical strategy for AI engineers in the era of deep learning. Retrieved from https://www.mlyear ning.org.

Sagiroglu, S. and Sinanc, D. (2013). Big data: A review. In *2013 IEEE International Conference on Collaboration Technologies and Systems (CTS)*, 42–47.

14.9 Sample Questions

Please select the most appropriate response.

Question 1

Which of the following belongs to big data processing process?

  I. Data acquisitions
 II. Data analysis and visualization
III. Data display

(a) I and II
(b) II and III
(c) I, II, and III

Question 2

Which of the following represents the correct order of the big data processing process?

(a) Data acquisitions—Data storage—Data processing—Data Analysis and Visualization—Data display

(b) Data storage—Data acquisitions—Data processing—Data Analysis and Visualization—Data display

(c) Data acquisitions—Data processing—Data storage—Data Analysis and Visualization—Data display

Question 3

Which of the following is not one of the three levels of Artificial Intelligence?

(a) Computing intelligence
(b) Vision intelligence
(c) Cognitive intelligence

Question 4

Which of the following are areas that we can apply artificial intelligence to?

I. Education, smart city, wearable smart device
II. Medical, security, autonomous driving
III. Build computing infrastructure

(a) I and II
(b) II and III
(c) I, II, and III

Question 5

Which of the following is not a characteristic of cloud computing?

(a) Low reliability
(b) Pay on demand
(c) Resources sharing

Solutions

Question 1

Solution: Option **c** is correct.

The complete big data processing process is mainly divided into five steps: data acquisition, data storage, and management, data processing, data analysis and visualization, data display.

Question 2

Solution: Option **a** is correct.

Question 3

Solution: Option **b** is correct.

It should be perception Intelligence.

Question 4

Solution: Option **a** is correct.

Education, smart city, wearable smart device, medical, security, and autonomous driving are all possible areas to apply AI for. Building computing infrastructure is not so relevant.

Question 5

Solution: Option **a** is correct.

Should be high reliability.

PART V

Blockchain, Cryptocurrency, and Investment

Chapter 15

Blockchain Technology

This section describes the differences between blockchain, distributed ledger technologies, and distributed databases from several aspects, including the adversarial and underlying data structures. It also examines various components in a blockchain system and details the characteristics of different blockchain types suited for different applications.

15.1 Learning Objectives

- Distinguish blockchain, distributed database, and distributed ledger technology.
- Examine the components in a blockchain system.
- Appraise the characteristics of different types of blockchain.

15.2 Main Takeaways

15.2.1 *Main Points*

- While the term "Blockchain" is often used interchangeably with "DLT", it is important to remember that some DLTs do not use a "chain of blocks" model.
- There are three types of blockchain: private blockchain, public blockchain, and consortium blockchain.
- Blockchain design is by nature immutable. This means that the information stored on the blockchain cannot be tampered with.

It achieves so via cryptographically binding a series of transactions into a sequentially ordered chain of blocks.

15.2.2 Main Terms

- **Blockchain:** Blockchain refers to a distributed record-keeping system that is append-only and secured by cryptography and consensus protocols.
- **Peer-to-Peer (P2P) Network:** Network that facilitates interaction, discovery, and data sharing between elements of the network. Simplistically speaking, each element represents a computer. Two or more computers can form a P2P network, allowing them to bypass a server and start sharing information.
- **Consensus mechanism:** The consensus mechanism is an algorithm that allows nodes in a decentralized system to determine which transaction is valid, how transactions are ordered, etc.

15.3 Introduction to Blockchain

15.3.1 Centralized vs Distributed Database

This section introduces both centralized and distributed database systems (Tupper, 2011). The first distinction is the location of components. All components of a centralized database are located on a single computer site. These components include the data itself, the distributed database management system (DDBMS), and other storage media needed to sustain the database's orderly operation. The centralization of all components into a single location is what gives the system its name. The second point of differentiation here is the mode of access. Under a centralized database, the access is provided by a combination of intelligent workstations and remote access terminals. For the remote access terminals, they utilize directed communications links.

Recently, a new technology had surfaced, allowing the data and process to be distributed across a large geographical area while being linked together by a communication network. This is known as a distributed system.

The first key difference between a distributed database vs a centralized database lies in their location. As the name suggests, a distributed system is no longer confined to a single computer site. On their own, these networks are considered as independent, intelligent communications computers and communication mechanisms. The distributed form of the communication system is vital in today's world. Due to globalization, we often experience huge geographical barriers, especially between different countries. These barriers are overcome with a distributed system as we can now link several fragmented sites located in separated regions.

Upon first glance, the user's view layers and the conceptual layer of a distributed model are almost identical to a traditional database design. The similarity lies with their coherency of data. This coherency of data within a distributed system is vital. Under the distributed system, the user is cut off from the physical view. As such, the importance of having a coherent link within the database becomes paramount. However, down at the physical level, things start to differ drastically. The key difference is caused by the distributed nature of the database. What this means is that the data is distributed across many sites, at many different machines. In terms of format or data structure, each site might hold this information differently.

Another key principle is to locate the data structure within a distributed system strategically. This strategic placement design reduces the chance of distributed joints from being formed to reduce the high overhead cost imposed by such joints. Sometimes, the cost can be further reduced by replicating data instead of fragmenting them. The decision to choose replication over fragmentation depends on the type of database it is, whether it is read-only or transactional.

However, despite the distinctions, there are also issues unique to the distributed system compared to the centralized systems. The main root of these problems is the high complexity of the distributed system. The distributed system is complex in terms of its physical and software components. There is also the complication from the network component, which had since integrated under the DDBMS. One such problem lies with the synchronization of the distributed information. Multiple copies of data need to be retained as reference data, and they need to be synchronized with each other. To achieve such consistency, a concurrency management mechanism needs to be put in place. These measures greatly extended the complexity of a distributed database as compared to a centralized one.

15.3.2 *Blockchain, Distributed Database, and Distributed Ledger Technology*

The term "Blockchain" originally describes the distributed record-keeping system used by the Bitcoin protocol. However, today, the term generally refers to any Distributed Ledger Technology (DLT) that is inspired by Bitcoin's blockchain design.

Blockchain refers to a distributed record-keeping system that is append-only and secured using cryptography and consensus protocols. While the term "Blockchain" is often used interchangeably with "DLT", it is important to remember that some DLTs do not use a "chain of blocks" model. The following describes the differences between the terms "distributed database", "DLT", and "Blockchain" (Infocomm Media Development Authority, 2020).

- *Distributed databases*: Databases that operate without centralized control. Each piece of data is copied across various nodes and devices. Together, these decentralized pieces of data are combined to maintain a holistic view of the entire database. Organizations normally use distributed databases with offices in different geographical locations.
- *Distributed ledgers*: The design takes into consideration the necessary defenses against an adversarial threat. By doing so, the distributed ledger can reduce malicious nodes within the network and are tolerant towards Byzantine faults.
- *Blockchains*: A derivative of distributed ledgers. A key distinction of blockchain is the use of blocks to bundle transactions and/or facilitate data dissemination to all blockchain participants.

Users (or nodes) in distributed databases are assumed to be trusted and will collaborate and perform database replication and updates honestly. However, distributed ledgers assume the presence of users who may perform malicious acts such as attempts to modify, delete, or add information to the ledger or simply refuse to work or collaborate. Blockchains assume the same adversarial threat model as distributed ledgers and use a "chain of cryptographically-linked blocks" as its underlying data structure. Figure 15.1 depicts this relationship.

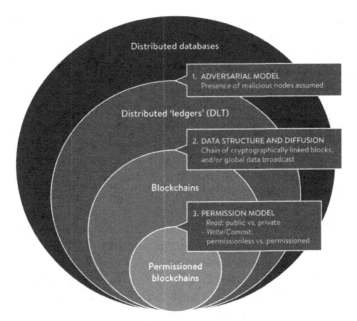

Figure 15.1: Blockchain, distributed ledger technology, and distributed database.
Source: Hileman and Rauchs (2017).

15.3.3 *Definition of Blockchain*

Regarding the NISTIR 8202 Blockchain Technology Overview document drafted by the National Institute of Standards and Technology, US Department of Commerce, blockchain is defined as the following:

> Blockchains are distributed digital ledgers of cryptographically signed transactions that are grouped into blocks. Each block is cryptographically linked to the previous one (making it tamper-evident) after validation and undergoing a consensus decision. As new blocks are added, older blocks become more difficult to modify (creating tamper resistance). New blocks are replicated across copies of the ledger within the network, and any conflicts are resolved automatically using established rules.
>
> (Yaga *et al.*, 2018)

15.3.4 *Blockchain Components*

A blockchain typically has the following components:

- *Wallet*: This typically refers to the devices that individuals and businesses use to access blockchain-based services. Digital wallets can either be (i) software wallets that are installed on the users' computers or mobile devices or (ii) hardware wallets that are essentially hardware devices protected by a secure chip (or equivalent). There are also cold wallets that are entirely offline wallets used for cold storage of cryptocurrencies and other digital assets on the blockchain and only connect to the Internet to send and receive such digital assets. Another variant is paper wallets, which are just pieces of paper with a private key written.

- *Nodes*: Generally, nodes are entities or computers that perform operations on the blockchain system. A node can be a normal user or a validator. Normal users create transactions through the use of wallets and use the blockchain as a platform to perform certain operations. Validators collect, validate, and add transactions to the blockchain. A transaction recorded on the blockchain is considered confirmed, although most public blockchain provides only probabilistic confirmation.

- *Transactions*: Depending on the actual blockchain system, a transaction is a data structure created by nodes and may contain information such as the sender, the receiver, the amount to transfer, and arbitrary data. Most public blockchain provides probabilistic confirmation of transactions. In other words, a transaction is considered "confirmed", that is, impossible to revert, only after a certain number of blocks are appended to the block containing the transaction. The time needed is called time to finality (Buterin, 2016).

- *Blocks*: A data structure on the blockchain that contains a list of transactions in the body and other data such as block number, timestamp, the hash value of the previous block, etc., in the header.

- *Cryptography*: Cryptography is considered a form of technique. The most prominent one discussed is the "one-way hash function". The one-way hash function generates a unique identifier for the block by calculating a hash value on the block header. Conceptually, each block stores the block header hash value of its preceding block. This design creates a link that joins all the blocks together, known as a cryptographic linkage. This linkage forms the basis behind the integrity protection of the blockchain, allowing blockchain to be resistant to tampering.

- *Peer-to-Peer (P2P) Network*: Network that facilitates interaction, discovery, and data sharing between elements of the network. Simplistically speaking, each element represents a computer. A combination of two or more computers forms a P2P network which allows these computers to bypass a server and start sharing information.
- *Consensus mechanism*: The consensus mechanism is an algorithm that allows nodes in a decentralized system to determine which transaction is valid, how transactions are ordered etc. I.
- *Validity rules*: A universally adopted set of regulations within the network. This may include rules which determine the validity of transactions or how the ledger may be updated.

At this juncture, it is important to stress the integrity of a blockchain design. Blockchain design is by nature immutable. This means that the information stored on the blockchain itself cannot be tampered with. It achieves so via cryptographically binding a series of transactions into a sequentially ordered chain of blocks. Hence, the name "blockchain" is derived. The cryptographic technique used is hash functions. During the process, what happens is that whenever a new block is added, its cryptographic hash value is validated against the previous block. Only upon successful validation can the new block be added to the end of the chain permanently. This process of validation is performed by other validating nodes. These nodes represent the other participants within the system, and they will offer validation through the use of digital signatures.

The system efficiently utilizes both the hash function and digital signature to create the chain of validated and yet, immutable records of blockchains. Any attempt to subvert the system or to change the ledger data will be economically infeasible. The economic infeasibility of subversion is possible due to the infeasibility of breaking the cryptographic techniques and the consensus algorithm's resiliency that requires at least half, or 50% of the nodes in the blockchain to be trustworthy. Hence, for an attacker to succeed in his attack to subvert the system, 51% of the nodes must be controlled by the attacker. This action by itself requires a large amount of computing power, which carries an extremely high-cost. Hence, attackers are disincentivized from attacking the blockchain (Lee and Low, 2018).

Application Layer	• Various application scenarios and cases
Contractive Layer	• Script codes, algorithmic mechanisms, smart contracts
Incentive Layer	• Issuance mechanism, distribution mechanism
Consensus Layer	• Consensus algorithm
Network Layer	• Distributed networking mechanism, data dissemination mechanism, data verification mechanism
Data Layer	• Data block, time stamping basic data, basic algorithms

Figure 15.2: Blockchain layers.

15.3.5 *Six-layers of Blockchain*

Wanxiang Blockchain Labs decomposes blockchain into six layers (Wanxiang, 2017), namely the data layer, network layer, consensus layer, incentive, contract, and application layer. A brief overview of each component can be found in Figure 15.2.

15.3.6 *Types of Blockchain*

Table 15.1 elaborates the main types of blockchains, segmented by permission model (Infocomm Media Development Authority, 2020).

Blockchains are highly use-case-specific technologies. Both types of blockchains—private and public—are designed to provide shared record-keeping access driven by participants' consensus and the option of running smart contracts. Otherwise, private and public blockchains have very little in common (Buterin, 2015).

Public blockchain assumes that anyone from the public Internet can join or leave the blockchain network without providing any forms of identification or asking for permission. For this to work, the design assumes that altruism does not exist and all actors are driven by their incentives. As a result, the consensus mechanism in a public blockchain should adopt the crypto-economics concept to incentivize collaboration and disincentivize cheating. An example of such a consensus mechanism is the proof-of-work (PoW) in Bitcoin. PoW rewards miners with tokens that can be traded in the real-world

Table 15.1: Different types of blockchains, segmented by permission model.

			Point of comparison		
			Read	Write	Validate
Blockchain types	Open blockchain	Public permissionless	Open access to all	All	All
		Public permissioned	Open access to all	Authorized personnel	All or specific individuals within the group of authorized personnel
	Closed blockchain	Consortium	Access is restricted to an approved list of participants	Authorized personnel	All or specific individuals within the group of authorized personnel
		Private permissioned (Enterprise)	Entirely private or under strict limitation to authorized nodes only	Only the operator of the network are allowed	Only the operator of the network is allowed

Source: Hileman and Rauchs (2017).

for value. It creates a disincentive for any user to subvert the network because the cost of real-world electricity investment outweighs any gains that can be derived from it.

Private blockchain assumes that all network participants are known and trusted and belong to a controlled membership. These participants can be individuals, such as employees of organizations like companies or departments within companies. Usually, the participants will draft physical, legal contracts and agreements to regulate their behavior. They are subjected to human rules and regulations and liabilities and obligations that are enforceable in the real-world legally. Therefore, a private blockchain can adopt a voting-based consensus mechanism instead of a computational-intensive consensus mechanism such as PoW.

Consequently, the solution can achieve better performance. Private blockchains may require different levels of access needs to be crafted for different roles of users. For example, to participate in consensus, to read or create a transaction, to execute a smart contract, one may be required to show a different set of permissions to be authorized. A central authority (or group) may also be required to perform permission review and approval.

The *consortium blockchain* is very similar to a hybrid state of blockchain, which is semi-private and semi-public. A distinct feature of this system is the inclusion of a controlled user group that operates across different organizations. For a more vivid description, it is a central collection of organizations, not a single authority that governs the blockchain in a distributed manner, much like a "federation" political system. In practice, consortium blockchain is used actively in cross-disciplinary efficiency. For instance, multiple stakeholders in the global supply chain form a consortium to share and trace goods' flow. This is vital for today's business world since cross-functional teams are often tasked to collaborate on a given task.

The main differentiating factor for a *permissioned* and *permissionless* model is whether participants are "authorized" to join the blockchain and perform write and/or validate operations. For instance, Bitcoin is a permissionless model because anyone can join and perform write and validation operations as long as they have a (relatively powerful) workstation with an Internet connection. Blockchain systems adopting Proof-of-Stake (PoS) require one to prove that a specific amount of deposit has been staked before one is allowed to perform write and validation operations on the blockchain. These systems adopt a permissioned model. Both permissioned and permissionless models have their pros and cons. The adoption of one model is subjective to the actual business use case, company policies, and even regulations, among other factors.

15.4   Blockchain Characteristics and Applications

This section looks at how the crypto-economics concept in blockchain governs nodes in different types of blockchain. The design thinking behind blockchain gives rise to many salient characteristics that make blockchain technology suitable for some industry applications when used correctly. This section will also describe how blockchain evolves from a decentralized peer-to-peer payment system as described by Satoshi Nakamoto to a technology that can be adopted as an industrial application.

15.4.1 *Learning Objectives*

- Appraise the crypto-economics concept in blockchain design.
- Explain the properties of a blockchain system.
- Examine the evolution of blockchain and explore potential applications for blockchain.

15.4.2 *Main Takeaways*

15.4.2.1 *Main Points*

- The first generation of blockchain starts with Bitcoin and cryptocurrencies. The second generation emerges with the development of Ethereum and smart contracts. The third generation of blockchain is under development, attempting to solve technology challenges, such as performance and scalability, interoperability, and privacy issues in the blockchain.
- A blockchain system provides a single version of the truth in a distributed manner. Blockchain is commonly considered as an immutable and ordered record of transactions. It leverages the salient features of cryptographic techniques and distributed consensus algorithm to provide a transparent, accountable, and synchronized "ledger" that is maintained by a group of mutually distrusting users across a geographically distributed network.

15.4.2.2 *Main Terms*

- **Crypto-economics:** The concept of crypto-economics is vital in a blockchain application. The term "crypto-economics" consists of two parts, "crypto" for cryptography and "economics" for economic incentives.
- **Ethereum:** A global decentralized application platform used for the development and operation of smart contracts.

- **Turing Completeness:** A mathematical concept and a measure of the computability of a programming language. A non-Turing Complete language means that the language is designed without complex constructs such as loops and conditions, limiting its ability to create general-purpose programs.

15.4.3   *Crypto-Economics*

The concept of crypto-economics is vital in a blockchain application. According to Vitalik Buterin, the term "crypto-economics" is made up of two words, namely "crypto" for cryptography and "economics" for economic incentives (Buterin, 2017). In general, cryptography is used in blockchain to prove properties about messages that happened in the past and to prevent dishonest users from attempting to cheat. Economic incentives are defined inside the blockchain system to encourage or incentivize desirable properties to hold into the future; it is used to reward honest users for up-keeping the integrity of the ledger and be part of a punishment system if users "misbehave".

- *Crypto-economic security margin*: An amount of money X such that you can prove "either a given guarantee G is satisfied, or those at fault for violating G are poorer than they otherwise would have been by at least X".
- *Crypto-economic proof*: A message signed by an actor can be interpreted as "I certify that either P is true, or I suffer an economic loss of size X".

15.4.3.1   *Example: PoW*

In the Bitcoin blockchain, new transactions are grouped into blocks to be added to the end of the current chain of blocks. The process is called mining, and the node (or user) that does the mining is known as the miner. Miners get a certain amount of bitcoins for every block successfully mined, and they can also collect the transaction fees of the transactions they include in that block. The total amount of bitcoin is fixed to 21 million, and the reward of mining will halve every 210,000 blocks mined (roughly every four years) (Bitcoin Wiki, 2020). The beginning reward is 50 BTC for every block, and the reward for one block now is 6.25 BTC (since May 2020). It is projected that if the mining power remained constant, the last Bitcoin would be mined in 2140.

The Bitcoin network's consensus mechanism is PoW and determines when a block could be successfully mined. The PoW consensus mechanism is designed to require all nodes of the network to compete for the reward. This competition involves searching for a nonce (a number that can be used only once in cryptography) by sheer brute-force use of processing power. The resulting block header hash[1] value is lower than the target value set by the network.

Block header hash value of the current block is included in the next block when the next block is formed. This creates a (cryptographic) "chain" of blocks. If a malicious node attempts to change transactions in one block, the block header hash value will change; this change will "propagate" to the next block and result in changes in the block header hash values of all subsequent blocks.

To this end, note that the usage of hash functions and block header hash value to form a cryptographic chain of blocks is not unique to Bitcoin. Most blockchains such as Ethereum, Litecoin, Zcash, and many others adopt a similar approach. They differ mainly in terms of the data that a block contains, the transaction structure, the rules that nodes in the system follow to determine when a block could be successfully mined. Forming a cryptographic chain of blocks makes modifications difficult. Any minor change to transactions in one block will cause the modification effect to spill over and be reflected in all subsequent blocks. An excellent demo on how modification effect propagates can be found on Anders Brownworth's page.[2]

Since it is possible for two or more different nodes to find a possible solution for the PoW algorithm almost simultaneously, the blockchain may fork temporarily (Schär, 2020), which allows parallel chains to co-exist temporarily. When such an event occurs, the nodes will pick the longest chain for the next append. Hence, for an attacker to subvert the network, it must compete for the right

---

[1]Hash function is a cryptographic technique that resembles a fingerprint generator. The hash value of a block (i.e., block hash value) can be regarded as a unique fingerprint (i.e., serial number) for that particular block.

[2]Brownworth, A. (n.d). Blockchain Demo. Retrieved from https://andersbrownworth.com/blockchain/blockchain.

to add a block and compete to produce the longest chain. By making it economically infeasible for malicious actors to spawn multiple nodes to game the system (also known as Sybil attack as detained in Matt (2018)), the bitcoin system disincentivizes bad behaviors such as attacks.

Note that in networks that utilize PoW, the more computational power a node has, the higher is its chances of finding the nonce that results in a block hash value lower than the target value set by the network. Nodes spend electricity and computation power in this process, but they are incentivized to do so because the winning node gets a block reward along with transaction fees. Also, they are incentivized to add blocks containing only valid transactions (i.e., not double-spending) because otherwise, their produced block is invalid and would be rejected by other honest miners; they would then have wasted electricity and computation power for nothing.

15.4.3.2 *Example: PoS*

PoS networks usually operate independently from miners. What these systems rely on were validators who carry voting power. For these validators, the amount of voting power they hold is proportionate to the amount of cryptocurrency (stake) they hold on the network. The rationale is that the more stake a validator holds, the less likely it is for the validator to attack the system—the effect of which devalues the stake in hand. Hence, the name "PoW" is derived from here.

This design has its own set of implications. Instead of using computational power, the validators place their funds as a deposit to qualify themselves the chance to add in new blocks. The higher the stake, the higher the probability of successfully adding their block. The validators will only receive the transaction fees for the newly added block if they are chosen and the transactions they propose are validated.

Initially, early PoS networks suffer from nothing-at-stake problems (Martinrz, 2018). Back in the initial implementations of PoS networks, for example, Peercoin, rewards are offered for producing valid blocks. However, no penalties were offered if users simultaneously create blocks on different chains. This penalty system incentivizes rational validators[3] to vote for multiple blocks at one go in an attempt to extract information from them all. As a result, the validators will not keep working on a single chain but rather add alternate blocks to multiple chains to outsmart the system's rules. This deprives the network from ever reaching a consensus on which chain is the "true one". This issue is known as the nothing-at-stake problem.

To overcome this issue, a sufficient penalty must be imposed to offer the right economic incentives. To do this, a penalty in terms of "security deposits" is imposed. Misbehaving nodes would forfeit some or all of their deposits. This countermeasure disincentives validators from adding invalid transactions or working on multiple chains at one go.

15.4.4 *Blockchain Properties*

15.4.4.1 *Immutability of Record*

The cryptographic linkages between blocks in the blockchain make data stored in the blockchain tamper-resistant. Information on the blockchain can only be "updated" but not amended. For instance, once a transaction is written into the blockchain, it cannot be deleted. Should an error be made, another transaction can be written to update the transaction's state, rectifying the error. In this case, the trail of transactions is preserved. This feature imparts confidence in the provenance of the value being transacted and enhances fraud detection.

15.4.4.2 *Single Version of Truth*

A blockchain system provides a single version of the truth in a distributed manner. Blockchain is commonly considered as an immutable and ordered record of transactions. It leveraged the salient features of cryptographic techniques and distributed consensus algorithm to provide a transparent, accountable, and synchronized

---

[3]Rational validators do this not to subvert the network, but rather, to maximize their own self-interest. Only the altruistic validators will forgo such interest and work continuously to benefit the network and abide by the rules.

"ledger" that is maintained by a group of mutually distrusting users across a geographically distributed network. For a block to be added to the blockchain, at least 51% of the validating nodes must reach a consensus on the block's validity and then update their ledger with the new block. Validators are incentivized to update to continue working on the next block and compete for the mining reward. Thus, a single version of truth can be provided in the absence of a centralized party.

15.4.4.3   *Peer-to-Peer (P2P) Data Transmission*

The peer-to-peer (P2P) network forms the backbone of the technology behind blockchain. A blockchain is a form of a distributed ledger. This means that instead of relying on a central authority, every node within the blockchain is empowered to communicate directly. Hence, authority or the power to communicate is distributed to each of the nodes in a blockchain. Since there is no central authority, and everyone exercises equal rights, the transaction within a blockchain would be made known in a P2P manner. This equitable structure implies that this transaction needs to be validated by all the blockchain nodes, using a fixed set of rules. Once validated, the transaction can then be added to the copy of the "ledger" maintained by each node. As equitable as it seems, this mechanism is not always viable. There could be potential delay or congestion which hinders nodes from receiving the same version of the blockchain. To overcome this error, each node keeps the state's higher scoring version. This means that whenever each node receives a state of a higher score, it will make changes to its ledger, and on top of that, it made this known to the other peers in the blockchain. This process made the verification of transactions easier. A node can simply query its nearby peers for their score. Upon sufficient cross-referencing, a node can then be aware of its scoring (Lee and Low, 2018).

15.4.4.4   *Disintermediation of Trust*

There is no central party in the network, mitigating against third-party risk. Many of the existing applications favor the solution of employing a trusted third-party to solve problems such as privacy, integrity, or detecting a breach of contract in business relationships. However, the reliance on a trusted third-party presents a single point of failure, and it is also a security loophole in the system design. In the blockchain, the immutable ledger together with the algorithmic consensus process becomes the trust agent and provides an infrastructure for transactions where intermediaries previously broker trust. Also, the smart contract enables the automation of business logic. As long as business terms, such as the conditions for payment, are agreed between two or more transacting parties, the logic can be coded in a smart contract that will automatically execute the agreed-upon terms when conditions are satisfied (Miller *et al.*, 2019). Research on smart contracts and their applications are always intertwined with whether a smart contract constitutes a legal contract. From the legal perspective, the challenges with smart contract adoption include (Levi and Lipton, 2018) the following:

- How can non-technical parties negotiate, draft, and adjudicate smart contracts?
- How does a smart contract extract information that is not available on the blockchain (i.e., off-chain data)?
- What is the "final" agreement between the transacting parties?
- Is the automated nature of smart contracts aligned with how businesses operate in the real-world?
- Which governing law should apply if users of the smart contract are from all over the world?

15.4.4.5 *Transparency and Traceability*

For public blockchains, every transaction made will be broadcast to the entire network without exception. After that, validator nodes will validate the transaction and put them into the block they are working on, later broadcast to other nodes. Every node has access to all transactions made in the blockchain because it is distributed and open-accessed.

15.4.4.6 *Scalability and Performance*

Distributed consensus protocols might have several advantages. However, it cannot simultaneously satisfy the requirement of scalability and performance. Indeed, a compromise has to be made since the level of distribution of a consensus protocol is inversely proportional to scalability and performance. This implies that high performance confers low scalability, vice versa (Vukolić, 2015).

A mechanism that caps performance is the target value. Under Bitcoin blockchain's PoW consensus, a fixed target value will be imposed based on the current network's collective computation power. The Bitcoin network requires that one block be produced around every 10 minutes. As such, the target value will be periodically adjusted (every two weeks) based on past data. If for the past two weeks, or 2016 blocks, the blocks are produced in less than 10 minutes on average, it means that the network's collective computation power is higher. Thus the target value will be fixed to a smaller value so that the time taken to produce a new block is adjusted back to 10 minutes on average.[4]

Therefore, a Bitcoin transaction needs to wait for six blocks before its confirmation, which is referred to as the six-block confirmation rule. This is known as the Bitcoin blockchain's "six-block confirmation rule". This rule is what gives PoW its bottleneck in terms of performance. Due to this constraint, the blockchain has a consensus latency of 60 minutes and a throughput of seven transactions per second (TPS) based on a block size of 1 MB. This bottleneck is what constrains the transactional speed of bitcoin compared to the credit card.

In comparison, bitcoin transaction falls behind severely from global credit card transaction average throughput (VISA, 2015) of 2000 TPS (with a peak of 56,000 TPS). However, Bitcoin has other advantages in terms of scalability. Bitcoin is far more scalable, and this scalability increases with the number of nodes joining the network. A higher number of users will also enhance the security of the network as well.

Conversely, the performance could be scaled up as well but with a compromise in terms of scalability. Conventional consensus protocol, for instance, PBFT, is usually designed to fit small-scale tasks. A typical consensus protocol could only handle file systems or databases of 10–20 nodes based on State Machine Replications. They are yet unproven to work on a much larger scale system such as Bitcoin.

---

[4]Blockchain.com. (n.d). Network difficulty. Retrieved from https://www.blockc hain.com/charts/difficulty.

This is because these conventional consensus protocols require the nodes to be authenticated and ahead of time while denying any nodes to enter and leave without permission. This design allows the consensus protocol to achieve high performance as there is a small existing group of authenticated individuals validating the transactions. Usually, transactions can occur at a rapid rate of few tens of thousands of transactions per second and are constrained by the network latencies. However, this design hinders scalability drastically. Since all nodes need to take part in the approval-making process, increasing the number of nodes would increase the time and resources required to facilitate such an approval process. In place of such constraint, conventional consensus protocol can only scale-up while ensuring the nodes are within close physical proximity (Lai and Lee, 2018).

15.4.4.7 *User Identity and Authentication*

Private blockchain has regulated or permissioned access. As such, the identity of users and their respective authentication becomes important in private blockchain as compared to public ones. However, a private blockchain is susceptible to Byzantine faults, and this effect will be more severe when the blockchain is unable to identify the erring nodes from the message. According to Lamport, some necessary conditions are necessary to reduce these Byzantine faults. For instance, consensus is allowed to converge for any number of Byzantine nodes if they are authenticated. The messages are also made unforgeable. These conditions are necessary to allow solutions that can tolerate more than 33% of Byzantine nodes (Lai and Lee, 2018).

15.4.4.8 *Smart Contracts*

Just like blockchain, there is no standard definition available for Smart Contracts. Ethereum popularizes the term smart contract and calls it a "mechanism involving digital assets and two or more parties, where some or all of the parties put assets in and assets are automatically redistributed among those parties according to a formula based on certain data that is not known at the time the contract is initiated". What Ethereum does is to equip developers with their own set of Turing Complete language known as Solidity. This ability is equivalent to offering a globally distributed computing platform (Lai and Lee, 2018). The multichain developers provided a comprehensive overview on how smart contracts on different platforms (i.e., Hyperledger Fabric, Multichain, Ethereum, and Corda) address the key challenges on (1) How are transaction rules represented? (2) How is code executed deterministically? and (3) How are conflicts prevented? (Greenspan, 2018).

15.4.5   *Blockchain Applications*

Since the publication of the Bitcoin paper in 2009, introducing blockchain technology to the world, the application of blockchain has evolved and diversified. The first generation of blockchain in the 2010s revolved around Bitcoin and cryptocurrencies. With the development of Ethereum and smart contracts, the application of blockchain extended beyond the finance sector and has since gained recognition for its potential to transform various industries. However, technology challenges, such as performance and scalability, interoperability, and privacy remain, and the third generation of blockchain is under development.

15.4.5.1   *Blockchain 1.0*

The representative Blockchain 1.0 system is Bitcoin. Bitcoin (uppercase "B") generally refers to the protocol and the network based on a paper published by Satoshi Nakamoto in 2008 (Nakamoto, 2008). The bitcoin (lowercase "b") cryptocurrency was released in January 2009 that is implemented with the Bitcoin protocol. The term blockchain refers to the underlying technology used in Bitcoin because of its block and chain architecture. The Bitcoin blockchain is a distributed public ledger where bitcoin transactions are recorded. Each block in the Bitcoin blockchain contains transaction data, and numerous blocks of data are chained together by cryptography.

Bitcoin is a digital currency relying on a peer-to-peer decentralized network. Centralized currency is issued by a central bank and traded through certain institutions like most traditional currency, while bitcoin is a kind of decentralized currencies in a peer-to-peer network with the generation and verification done by nodes (users) instead of a centralized third-party.

Before the occurrence of the Bitcoin protocol, the main challenge of digital currency is to solve the so-called double-spending problem.

Unlike any physical object, the digital currency appears in the form of data. It is thus possible to be copied and sent to two different recipients at the same time. This is not a difficult problem to solve when using a central authority that has a global view of the currencies in circulation, but decentralized platforms cannot resolve this issue easily. The difficult question to answer is: Which version of the ledger is the truth?

Bitcoin, however, was the first that successfully solved this problem in a fully decentralized manner. It achieves this using a combination of a distributed record-keeping system that is immutable and a consensus protocol that is designed with the crypto-economics concept. Instead of using a centralized authority to oversee and thwart malicious actions, public blockchains inspired by Bitcoin's design uses reward and penalizing mechanisms to incentivize good behaviors and disincentivize malicious acts. A distributed record-keeping system means that as long as more than 51% of the nodes in the blockchain system remain honest, the ledger can be trusted to be legit, that is, containing only legitimate transactions.

The Bitcoin blockchain is the main representation for Blockchain 1.0. Its main function is to facilitate decentralized peer-to-peer payments. The Bitcoin blockchain is completely transparent. In other words, the sender and receiver in every transaction are shown in the clear (although the real identity of the sender and receiver is concealed behind wallet addresses). Further, the amount transacted is also clear. Another class of cryptocurrency that adds a layer of privacy for such a decentralized peer-to-peer payment system is the privacy coin. The representative system for this class of blockchain is Zcash.

Among the pioneers of a public blockchain is Zcash.[5] Zcash is a significant case study as it represents the first public blockchain with a full privacy protection scheme. This is achieved via two mechanisms: encryption and zero-knowledge succinct non-interactive arguments of knowledge (zk-SNARKS) algorithm. Encryption is done on transaction data, while zk-SNARKs is used to ensure users' anonymity and the privacy of their transactions. When performing a transaction via Zcash, the value of the transaction can

---

[5]Zcash (2021). Retrieved from https://z.cash/.

either be transparent or shielded. These have different implications. A transparent transfer is similar in mechanism to Bitcoin, while shielded transactions are unique to Zcash. This is achieved by the unique design of Zcash. Their design includes two addresses, private (z-address) and transparent (t-address). This design empowered three forms of transactions, namely between two t-addresses, two z-addresses, or between a z-address and a t-address. Among these transactions, t-address to t-address is similar to Bitcoin transactions as it makes transaction parties and amount into public information. However, a transaction between two z-addresses is private. Both the transaction parties and the transaction amount are encrypted, in contrast to a Bitcoin transaction. Lastly, shielding transactions incorporate transactions performed from a t-address to a z-address. Reversely, a transaction that occurs from a z-address to a t-address will be known as a "deshielding" transaction (Lo *et al.*, 2018).

15.4.5.2   *Blockchain 2.0*

Ethereum is a global decentralized application platform used for the development and operation of smart contracts (Antonopoulos and Wood, 2018). It was launched amid much fanfare in July 2014 supported through an initial coin sale of 60 million ether valued at around $18 million.

While Bitcoin serves solely as a virtual currency primarily used for payment, Ethereum sought to perform more than just transfer cryptocurrencies between individuals and entities by incorporating smart contracts capabilities into its blockchain. Smart contracts allow the creation of decentralized applications that enable complex, peer-to-peer interactions between parties without relying on a trusted third-party. Simply put, smart contracts are capable of facilitating the automated exchange of assets, property, money, and anything of value transparently while removing any intermediaries in the process.

"Trust-less" stems from the fact that on public blockchains, users of the blockchain do not need to trust the other party or the smart contract because the blockchain by design prevents incorrect transactions. So oddly, "trust-less" means that trust is not needed. In other words, we can have absolute trust in the correct operation of a trust-less system, and we need not know or trust the counter-party.

All transactions that interact with the blockchain and any changes made to the state of an address or a smart contract can be publicly viewed on a ledger. Hence, the greatest value proposition of public blockchains like Ethereum is accountability.

The Bitcoin blockchain contains a basic form of a programming language called Script (O'Reilly, n.d.). Script expressions use reverse polish notation and are processed using stacks and postfix algorithms. The results arising from Script operations are limited to Boolean output only, that is, either true or false. The Script language used in Bitcoin is also known as a non-Turing Complete language. It is an elementary, stack-based programming language that processes the transactions on the Bitcoin blockchain.

Turing Completeness is a mathematical concept and a measure of the computability of a programming language.[6] A non-Turing Complete language means that the language is designed without complex constructs such as loops and conditions, limiting its ability to create general-purpose programs. In Bitcoin's case, this is on purpose as it avoids the risks of bad programming such as infinite loop from bringing down the entire system.

Ethereum began as a proposition for Bitcoin 2.0; it provides developers with its Turing Complete language called Solidity, which is essentially a global-distributed computing platform.

Table 15.2 summarizes the comparison between Bitcoin and Ethereum.

Given that Ethereum is the most well-known smart contract platform today, most use Ethereum's definition of a "smart contract" as "a mechanism involving digital assets and two or more parties, where some or all of the parties put assets in". The assets are automatically redistributed among those parties according to a formula based on certain data that is not known at the time the contract is initiated.

Although Bitcoin blockchain can also be used to create a smart contract, it will be very difficult to use. This is similar to writing software applications using calculators (which was what reverse polish notations were normally used for). Ethereum utilizes blockchain

---

[6]A good video introduction to the Turing Complete concept can be found at Computerphile's page at https://www.youtube.com/watch?v$=$dNR DvLACg5Q.

Table 15.2: A comparison between Bitcoin and Ethereum.

	Bitcoin	**Ethereum**
Token	bitcoin (BTC)	ether (ETH)
Blockchain	Bitcoin Blockchain	Ethereum Blockchain
Consensus	Proof-of-Work	Proof-of-Work/Proof-of-Stake (Scheduled switch from PoW to PoS is sometime in 2021)
Programming Language	Script	Solidity

technology to maintain both a decentralized payment network and store computer code that can be used to power tamper-proof decentralized financial contracts and applications. Ethereum applications and contracts are powered by ether, the Ethereum network's currency.

One way to look at smart contracts is to view them as a dis-intermediary of lawyers if Bitcoin can be described as a dis-intermediary of the central monetary system. Yet, this is not practical in the business world since legal entities are governed by laws, which require human interpretations and enforcements.

Today, smart contracts are largely coded with the Solidity programming language. However, Solidity is not the only programming language that supports smart contracts. Vyper, a Python 3 derived programming language, is also gaining popularity. In Hyperledger Fabric that similarly supports smart contract (or chaincode), the programming languages that can be used include Go, Javascript, or Java, SDKs in Node.js, Java, Go, REST and Python.

The logic of smart contracts is coded using the programming language of choice and is compiled down to bytecode or machine-readable code. Bytecode is then interpreted and executed by the Ethereum Virtual Machine (EVM). See Figure 15.3.

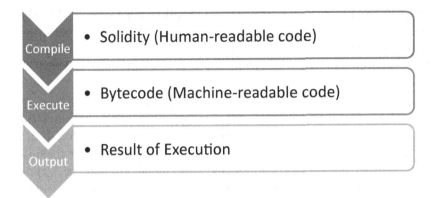

Figure 15.3: The process of compiling and executing a smart contract.

Before deploying a smart contract, it is fundamental to note that there are some differences between decentralized applications and traditional programming. Unlike traditional programming, where one can simply change the code and redeploy the program, this process is not as straightforward for smart contracts. Once a smart contract is deployed (i.e., published) to the Ethereum blockchain, it is immutable and cannot be edited anymore. Additionally, executing every line of code in a smart contract costs a fee, hence codes must be optimized before deployment. To assist developers with this process, test nets have been created to mimic the performance of an actual contract on the actual Ethereum blockchain or the mainnet. It is important to note that a test net cannot interact with the real blockchain transactions on the Ethereum blockchain.

It is also worth noting that most smart contracts deployed on public blockchains are visible to everyone. Shielded smart contracts, like privacy coins that work with shielded transactions and ledgers, are technically feasible.

15.4.5.3 *Blockchain 3.0*

With the development of blockchain and smart contracts, blockchain applications extended beyond the finance sector have since gained recognition for their potential to transform various industries. These applications leverage blockchain's properties to reduce the cost of trust and improve business efficiency. The United Nations Development Programme (UNDP) identified six areas that blockchain can do for the World Bank's Sustainable Development Goals. These areas are support financial inclusion, affordable and clean energy, responsible consumption and production, protect the environment, provide legal identity for all, and improve aid effectiveness.

However, technology challenges, such as performance and scalability, interoperability, security, and data privacy remain. As blockchain is adopted to solve real-world problems, it is inevitable for data or devices on the blockchain ecosystem to interface with other technologies such as Internet-of-Things (IoT), artificial intelligence, cryptographic protocols, payment gateways, and secure hardware. Studying the performance, scalability, security, and privacy of an ecosystem as such is crucial.

Interoperability and convergence are two key features of any working sociotechnical infrastructure. As information systems scale-up and the heterogeneity of users increases, there are risks and complexities in the actualization of interoperability and convergence. With interoperability and convergence, social scalability and mass adoption of blockchain in the industry can be achieved.

Blockchain 3.0 is a stage where the community attempts to fix these problems to remove the roadblock to mass adoption.

Bibliography

Antonopoulos, A. M. and Wood, G. (2018). *Mastering Ethereum*. O'Reilly Media, Inc., Sebastopol, CA.

Bitcoin Wiki. (2020). Controlled supply. Retrieved from https://en.bitcoin.it/wiki/Controlled_supply.

Blockchain.com. (n.d). Network Difficulty. Retrieved from https://www.blockchain.com/charts/difficulty.

Buterin, V. (2015). On Public and Private Blockchains. Ethereum Blog. Retrieved from https://blog.ethereum.org/2015/08/07/on-public-and-private-blockchains/

Buterin, V. (2016). On Settlement Finality. Ethereum Blog. Retrieved from https://blog.ethereum.org/2016/05/09/on-settlement-finality/.

Buterin, V. (2017). Introduction to Cryptoeconomics. Retrieved from https://2017.edcon.io/ppt/one/Vitalik%20Buterin_Introduction%20to%20Cryptoeconomics_EDCON.pdf.

Greenspan, G. (2018). Smart contract showdown: Hyperledger Fabric vs MultiChain vs Ethereum vs Corda. MultiChain. Retrieved from https://www.multichain.com/blog/2018/12/smart-contract-showdown/.

Hilcman, G. and Rauchs, M. (2017). Global blockchain benchmarking study. Cambridge Centre for Alternative Finance. Retrieved from https://assets.ey.com/content/dam/ey-sites/ey-com/en_gl/topics/emeia-financial-services/ey-global-blockchain-benchmarking-study-2017.pdf.

Infocomm Media Development Authority (2020). Artificial intelligence, data and blockchain in a Digital Economy. World Scientific. Singapore University of Social Sciences-World Scientific Future Economy Series: Volume 3. Retrieved from https://www.worldscientific.com/worldscibooks/10.1142/11787.

Lai, R. and Lee, D. (2018). Handbook of blockchain, digital finance, and inclusion (Chapter 7 Blockchain — From public to private), 145–177. Retrieved from https://www.sciencedirect.com/science/article/pii/B9780128122822000073.

Lee, D. K. C. and Low, L. (2018). *Inclusive FinTech: Blockchain, Cryptocurrency and ICO*. World Scientific, Singapore.

Levi, S. D. and Lipton, A. B. (2018). An introduction to smart contracts and their potential and inherent limitations. Harvard Law School Forum on Corporate Governance. Retrieved from https://corpgov.law.harvard.edu/2018/05/26/an-introduction-to-smart-contracts-and-their-potential-and-inherent-limitations/.

Lo, S. W, Wang, Y. and Lee, D. (2021). Blockchain and smart contracts: Design thinking and programming for FinTech. Singapore University of Social Sciences-World Scientific Future Economy Series: Volume 4. Retrieved from https://www.worldscientific.com/worldscibooks/10.1142/11919.

Martinrz, J. (2018). *Understanding Proof of Stake: The Nothing at Stake Theory*. Coinmonks. Retrieved from https://medium.com/coinmonks/understanding-proof-of-stake-the-nothing-at-stake-theory-1f0d71bc027.

Matt, B. (2018). Bitcoin's attack vectors: Sybil & eclipse attacks. Chainrift Research. Retrieved from https://medium.com/chainrift-research/bitcoins-attack-vectors-sybil-eclipse-attacks-d1b6679963e5.

Miller, D. *et al.* (2019). Blockchain: Opportunities for private enterprises in emerging markets. International Finance Corporation, World Bank Group. Retrieved from http://documents1.worldbank.org/curated/pt/260121548673898731/pdf/134063-WP-121278-2nd-edition-IFC-EMCompass-Blockchain-Report-PUBLIC.pdf.

O'Reilly (n.d). Chapter 5. Transactions. Retrieved from https://www.oreilly.com/library/view/mastering-bitcoin/9781491902639/ch05.html.

Nakamoto, S. (2008). Bitcoin: A peer-to-peer electronic cash system. Retrieved from https://bitcoin.org/bitcoin.pdf.

Schär, F. (2020). Blockchain forks: A formal classification framework and persistency analysis. Munich Personal RePEc Archive. Retrieved from h ttps://mpra.ub.uni-muenchen.de/101712/1/MPRA_paper_101712.pdf.

Tupper, C. (2011). *Data Architecture: From Zen to Reality.* Elsevier, Burlington, MA.

UNDP. (2020). Beyond bitcoin. UNDP |Beyond Bitcoin. https://feature.u ndp.org/beyond-bitcoin/.

VISA (2015). Visa fact sheet. Retrieved from: https://usa.visa.com/dam/ VCOM/global/about-visa/documents/aboutvisafactsheet.pdf.

Vukolić, M. (2015). The Quest for Scalable Blockchain Fabric: Proof-of-Work vs. BFT Replication. International Workshop on Open Problems in Network Security (iNetSec), Zurich, Switzerland, 112–125, ff10.1007/978-3-319-39028-4_9ff. ffhal-01445797f. Retrieved from https://hal.inria.fr/hal-01445797/document.

Wanxiang Blockchain Labs (2017). The digital migration for mankind. Retrieved from https://www.blockchainlabs.org/index_en.html.

Yaga, D. *et al.* (2018). Blockchain technology overview. National Institute of Standards and Technology, US Department of Commerce. Retrieved from https://csrc.nist.gov/CSRC/media/Publications/nistir/8202/dra ft/documents/nistir8202-draft.pdf.

15.5 Sample Question

Please select the most appropriate response.

Question 1
What does the EVM do?

(a) Validates transactions
(b) Executes smart contracts
(c) Secures Bitcoin

Question 2
What does "immutability of record" mean?

(a) Records on a public blockchain cannot be updated or amended but can be deleted
(b) Records on a public blockchain cannot be deleted
(c) Records on a public blockchain can only have their states updated

Question 3

Which of the following is not a feature of public permissioned blockchain?

(a) The data on the blockchain can be read by anyone
(b) Only authorized personnel can write data to the blockchain
(c) Anyone can validate transactions on the blockchain

Question 4

Which of the following is true about Proof-of-Stake?

(a) It requires validators to possess high computational power to successfully add a block
(b) It requires validators to place a deposit to qualify for the chance to add a block
(c) If a validator changes the transactions in one block, the effect of the change will "propagate" to all subsequent blocks

Question 5

Which of the following is true about Ethereum?

(a) The smart contract can be coded using Solidity, Java, and Python
(b) The Ethereum blockchain records both payment information and smart contract code(c) Unlike Bitcoin, the Ethereum blockchain has a much higher throughput comparable to Visa

Solutions

Question 1

Solution: Option **b** is correct.

Question 2

Solution: Option **c** is correct.

Question 3

Solution: Option **c** is correct.

Question 4

Solution: Option **b** is correct.

Option a refers to Proof-of-Work and option c applies to all blockchain designs that use the cryptographic hash to form the chain of blocks.

Question 5

Solution: Option **b** is correct.

Option a is incorrect because, at present, Ethereum's smart contract language is Solidity. Option c is incorrect because Ethereum, like Bitcoin, is a public blockchain, where anyone can join the network. Hence, the consensus protocol design limits public blockchains' throughput.

Chapter 16

Cryptography

Cryptography is a method of encrypting or decrypting information to conceal its real meaning with other technology and transmit information secretly. In computer science, it refers to utilizing an algorithm that is difficult to decipher, transform messages, and guarantee data security. The *Merriam-Webster*'s definition of cryptography is the "enciphering and deciphering of messages in secret code" but has recently added the definition "the computerized encoding and decoding of information" to encompass its more commonly known modern definition. The modern definition links cryptography with many more digital terms such as hashes and digital signatures. This section aims to describe a brief history of modern cryptography with a focus specifically on the hash function, public key infrastructure, and digital signature.

16.1 Learning Objectives

- Describe what cryptography is and explain the design rationale of hash function and digital signature.
- Examine how cryptographic techniques are used to secure public/private blockchains.

16.2   Main Takeaways

16.2.1   *Main Points*

- The security of cryptographic techniques is based on the intractability of solving difficult mathematical problems.
- Hash function and digital signature schemes output strings of text that appear random.
- Both public and private blockchains use cryptographic techniques extensively to provide user and data authentication, as well as to protect the integrity of the distributed ledger.
- Cryptographic techniques are also used in distributed consensus algorithms such as Proof-of-Work (PoW).
- The private key, public key, and wallet address generation process for one user are one-way and irreversible.

16.2.2   *Main Terms*

- **Hash function:** A cryptographic function with properties similar to a fingerprint function. It outputs a unique "fingerprint" for input data.
- **Digital signature:** A cryptographic technique that works with public and private keys and for users to prove ownership of wallet address.
- **Private and public keys:** A pair of keys that belong to a single user. The private key is used to "unlock" funds for the user to spend; the public key is like an account number for the user to receive funds.
- **Proof of Work:** The consensus mechanism in Bitcoin that uses hash function extensively.

16.3   Introduction to Cryptography

Classical cryptography has been around for millennia—the earliest known uses of codes have been recorded as far back as ancient Egypt. What made classical cryptography an art was the fact that there was little theory behind the construction or decryption of codes, and no systematic way of thinking about the requirements a secure code had to satisfy. Its purpose was primarily to achieve secrecy, and because of the great expense involved, its use was limited to governments and military organizations. Perhaps the most famous example is the Enigma machine invented by the Germans in the early 20th century and used by Nazi Germany to encrypt military communications during World War II.

The field of cryptography has evolved a lot since then. Unlike classical cryptography, modern cryptography is not just an art but a science and a mathematical discipline. Instead of ill-defined, intuitive notions of complexity or cleverness, the field now relies on rigorous proofs of security. We can think of modern cryptography as a suite of algorithms based on the intractability of difficult problems, which are problems that cannot be solved in a "reasonable amount of time". In other words, existing cryptographic algorithms are not perfect; they are simply computationally infeasible to break (Henderson, 2013).

Compared to classical cryptography, modern cryptography is much more pervasive. It now extends beyond secret communication to the protection of the user, as well as data at rest (in storage) and in transit (sent over a network), and it is integral to nearly all computer systems. Cryptography has been widely adopted in people's daily lives (e.g., sending e-mails or paying for a ride with our transportation card).

16.3.1 *Hash Function*

A hash function is a method of mapping an input of arbitrary length onto an output of a fixed length. This output is termed as a hash value ("hash" or "message digest"). Using programming terminology, the hash function, denoted as Hash(), takes as input a message m of arbitrary length and produces a hash value h of fixed length l:

$$h = \text{Hash}(m)$$

Examples of a hash function include (the depreciated) MD family, namely MD2, MD4, and MD5, the SHA-1 (broken by CWI Amsterdam and Google in 2017), SHA-2, SHA-3 family, BLAKE family, RIPEMD hash family, and other families. Different hash functions produce hash values of different length as shown in Figure 16.1 and many different factors such as security and performance, security

Hash Family	Members	Length
Message Digest (MD) [Depreciated]	• MD2 • MD4 • MD5 • MD6	• 128 bits • 128 bits • 128 bits • Variable*
Secure Hash Algorithm (SHA)	• SHA-0 • SHA-1 • SHA-2 • SHA-3 (formerly Keccak)	• 160 bits (retired) • 160 bits (retired) • Variable* • Variable*
RIPE Message Digest (RIPEMD)	• RIPEMD-128 • RIPEMD-160 • RIPEMD-256 • RIPEMD-320	• 128 bits • 160 bits (most popular) • 256 bits • 320 bits
BLAKE	• BLAKE-224 • BLAKE-256 • BLAKE-512 • BLAKE-384 • BLAKE2 • BLAKE3	• 224 bits • 256 bits • 512 bits • 384 bits • Variable* • Variable*

Figure 16.1: List of hash families and their members. Note that this list is not exhaustive.

Note: *Depending on the variant

objective, and length that affect the process of determining which to use.

This immutability is a result of the use of distributed ledger technology coupled with hash functions, which do not allow any modification to a block and its transactions to go undetected. Informally, a hash function possesses the properties shown in Figure 16.2.

At a very high-level, a hash function achieves these properties by taking in the input data and performs permutations and substitutions on the input data to destroy the statistical correlation between the input data and the hash value. As such, it is computationally infeasible for one to determine that the input data have given a hash value, and it is computationally infeasible for one to break the second pre-image resistant and collision-resistant properties of a hash function. To break these properties, a malicious user can do no better than performing a brute-force (or trial and error) by trying all possible input data and check on the resulting hash value.

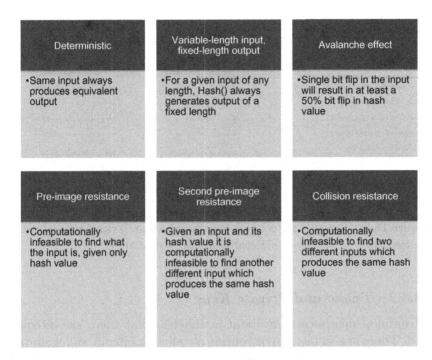

Figure 16.2: Properties of the hash function.

The main objective of a hash function is to integrity-protect data, although some applications use its one-way property to protect privacy. The hash function outputs a hash value that is unique to the input data. By securing the hash value and making it immutable (on a distributed ledger, for example), one can prove that the data is original and has not been altered as finding a collision in the hash value is computationally infeasible.

Hash functions are used in a myriad of applications to protect data integrity. For instance, they are used as integrity checksums for software downloaded from the Internet, as shown in Figure 16.3.

In the blockchain, it is used to generate a unique transaction "serial number" (i.e., transaction hash value) and to integrity-protect the blocks from modification. Bitcoin uses the SHA-256 and RIPEMD hash functions. The Ethereum blockchain uses Keccak-256, while Hyperledger Fabric uses SHA3 SHAKE 256 as its primary hash function.

Ghostscript is available under both an Open Source AGPL license and Commercial license. Please visit artifex.com/licensing/ to understand the differences in these licensing agreements, or to acquire a commercial license.

Platform/License	*AGPLv3* GNU Affero General Public License	**Artifex** Artifex Commercial License
Ghostscript 9.53.3 for Windows (32 bit)	Ghostscript AGPL Release	Ghostscript Commercial License
Ghostscript 9.53.3 for Windows (64 bit)	Ghostscript AGPL Release	Ghostscript Commercial License
Ghostscript 9.53.3 for Linux x86 (32 bit)	Ghostscript AGPL Release	Ghostscript Commercial License
Ghostscript 9.53.3 for Linux x86 (64 bit)	Ghostscript AGPL Release	Ghostscript Commercial License
Ghostscript 9.53.3 Source for all platforms	Ghostscript AGPL Release	Ghostscript Commercial License

NOTE: The Linux binaries are for testing/evaluation purposes only, they do not come as installers or installable packages.

MD5 Checksums
SHA512 Checksums

Figure 16.3: Example where a hash function is used to check software integrity. *Source*: https://www.ghostscript.com/download/gsdnld.html.

16.3.2   *Public and Private Keys*

A common misconception about a wallet is that there are bitcoins in it. There are, in fact, no bitcoins in a wallet. A bitcoin wallet simply reflects the bitcoin wallet address. Beyond that, it is a database of private keys which belonged to a user. Every user, besides having a private key, has a public key as well. The private key is unique to the owner and is pretty much kept in secret just like our bank account password. The public key, however, is something that anyone can see.

The relationship between private key, public key, and the wallet address is as follows:

1. Upon sign-up, a user's private key is generated at random.
2. The user's public key is computed from the private key using a predetermined algorithm. The algorithm computation is one-way in that if one has the private key, one can compute the corresponding public key but not vice versa. The one-way property is guaranteed by the infeasibility of our present computing power to solve specific difficult mathematical problems (such as the discrete logarithm problem[1]). Generally, to prove by contradiction—if our

---

[1]Flylib.com. (n.d). Discrete logarithms. Retrieved from https://flylib.com/boo ks/en/3.190.1.79/1/.

computer can solve the difficult mathematical problem, then it can also obtain the private key when given just the public key.

3. The relationship between the user's private key and public key is one-to-one. In other words, given a private key, there is one and only one corresponding public key.

4. Given the public key, a predetermined and one-way algorithm (e.g., hash function) is used to compute the user's wallet address.

Consequently, bitcoins within a wallet can only be unlocked by the corresponding private key, and only the owner of the wallet has the private key. Given the one-way and one-to-one relationships, no other users can deduce the private key given just the wallet address and/or public key, thereby no other users can unlock bitcoins in a wallet as long as the owner keeps the private key secure and secret.

16.3.2.1 *Bitcoin Example*

The Bitcoin private key is a 256-bit number generated in a truly random way. The phrase "truly random" means that the generation process takes parameters from sources that cannot be replicated nor imitated (such as a 15-second random trajectory of a cursor). This essentially means that if one loses (misplace or forgot) the private key, it is impossible to recover it, and as a result, bitcoins in the wallet corresponding to the private key are "lost" (i.e., cannot be unlocked, thus can no longer be spent).

The Bitcoin private key can be stored either as:

- A hexadecimal number (see Figure 16.4), or
- A Base-58 encoded string also known as Wallet Interchange Format (WiF), is shown in Figure 16.5.

The private key should be managed by the owner independently as disclosing the private key will grant others access to your currency,

Figure 16.4: Private key stored as a hexadecimal number.

Wallet Import Format (Base58Check)

VEEWgYhDhqWnNnDCXXjirJYXGDFPjH1B8v6hmcnj1kLXrkpxArmz7xXw

Figure 16.5: Private key stored in Wallet Interchange Format.

Note: Two encoding systems are used here: hexadecimal and Base58. Base58 is a binary-to-text encoding system used to represent large integers as alphanumeric text, introduced by Satoshi Nakamoto for use with Bitcoin. It consists of the twenty-six letters of the alphabet, both upper and lower case, and the numbers 0–9, while excluding 0 (zero), O (capital o), I (capital i), and l (lower case L) to avoid confusion to the human eyes. Figure 16.6 exemplifies how decimal numbers are represented by the two encoding systems:

Dec	Hex	Base58		Dec	Hex	Base58		Dec	Hex	Base58		Dec	Hex	Base58
0	0	1		15	F	G		30	1E	X		45	2D	n
1	1	2		16	10	H		31	1F	Y		46	2E	o
2	2	3		17	11	J		32	20	Z		47	2F	p
3	3	4		18	12	K		33	21	a		48	30	q
4	4	5		19	13	L		34	22	b		49	31	r
5	5	6		20	14	M		35	23	c		50	32	s
6	6	7		21	15	N		36	24	d		51	33	t
7	7	8		22	16	P		37	25	e		52	34	u
8	8	9		23	17	Q		38	26	f		53	35	v
9	9	A		24	18	R		39	27	g		54	36	w
10	A	B		25	19	S		40	28	h		55	37	x
11	B	C		26	1A	T		41	29	i		56	38	y
12	C	D		27	1B	U		42	2A	j		57	39	z
13	D	E		28	1C	V		43	2B	k				
14	E	F		29	1D	W		44	2C	m				

Figure 16.6: Representation of numbers in the two encoding systems.

assets, streams, and permissions, for instance, by Hardware Security Modules (HSM).

The Bitcoin public key is derived from the Bitcoin private key. The public key can be used by others to verify your ownership of the wallet without needing to know your private key.

A public key is represented by:

- A 65-byte hexadecimal number that starts with a 4 (old uncompressed format), or
- A 33-byte hexadecimal number that starts with a 2 or 3 (new compressed format)

The address of the Bitcoin wallet is derived from a public key. The wallet address is used

- to send you native currency or native assets[2] on the blockchain by others
- to verify your ownership to items recorded on the blockchain by others
- to verify your permissions by the blockchain.

A wallet address looks like this (see Figure 16.7). A Bitcoin wallet address may start with "1", "3", or "bc1", depending on the actual function.[3]

But it is stored internally like this with 25 bytes (see Figure 16.8).

Figure 16.9 shows the public key to wallet address generation for Bitcoin.

16.3.2.2 *Ethereum Example*

In contrast to Bitcoin, the Ethereum wallet address generation is very simple. Ethereum addresses are simply hexadecimal numbers, identifiers derived from the last 20 bytes of the Keccak-256 hash of the public key.

Figure 16.7: Wallet address.

---

[2]Native currency is the cryptocurrency on the underlying blockchain. For example, ether is the native currency of the Ethereum blockchain, but the ERC-20 tokens that use Ethereum blockchain are not the native currencies or assets because they depend on other blockchains.

[3]Wikipedia. (2020). Invoice address. Retrieved from https://en.bitcoin.it/wiki/Invoice_address.

Figure 16.8: Bitcoin wallet storage.

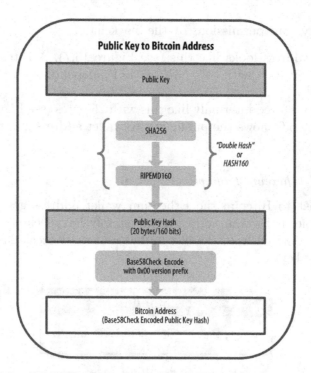

Figure 16.9: Public key to Bitcoin wallet address generation.
Source: Antonopoulos (2017).

16.3.3 *Digital Signature*

Digital signature is a pseudorandom digital string produced by
senders (signers) using his/her *private (signing) key* to prove the

authenticity (integrity protection and origin verification) of the information that the digital signature is computed on. The authenticity of the information can be verified by anyone using the sender (signer's) *public verification key.*

In the Bitcoin network, to do a transaction, the sender signs (note that the act of "signing" is essentially performing a mathematical computation using the data, private key, and other system parameters that are public information to the entire network) the data using his or her private key and sends the signed data together with the computed signature (a very large integer), and his or her corresponding public key to the receiver (the public key can be seen by everyone in the network).

The receiver can then verify whether the data is signed by the sender (or the sender is indeed the owner of that message) using the public key. As mentioned previously, the private and public keys are one-to-one and they possess a mathematical relationship. The act of "verifying" is simply another mathematical computation using the public key, the data, and the signature (computed using the private key). Due to the mathematical relationship between the private and public key, the "verifying" computation will output a number of either "1" (successful verification) or "0" (failed verification).

All in all, a *digital signature* is a mathematical scheme for verifying the authenticity of digital messages or documents. A valid digital signature, where the prerequisites are satisfied, gives a recipient very strong reason to believe that the message was created by a known sender (authentication), and that the message was not altered in transit (integrity). It also prevents the sender from denying that he/she has created this message (accountability) because only he/she possess the private key to compute the signature.

Digital signatures are a standard element of most cryptographic protocol suites, and are commonly used for software distribution, financial transactions, contract management software, and other cases where it is important to ascertain origin, detect forgery or tampering.

The *RSA digital signature scheme* applies the sender's private key to a message to generate a signature. The signature can then be verified by applying the corresponding public key to the message and the signature through the verification process, providing either a valid or invalid result (Kaliski, 2001).

Digital signature in blockchain is a way for the sender to prove that he or she has the authority to spend the bitcoin in a specific wallet. However, the RSA digital signature is not used in most blockchain systems. Instead, another variant of digital signature called the elliptic curve digital signature algorithm (ECDSA) is used. In cryptography, ECDSA offers a variant of the digital signature algorithm (DSA), which uses elliptic curve cryptography.

16.4 Cryptography and Blockchain

To recap, hash, or hashing, refers to a function that takes input data of any length and returns a value of fixed length. One of the properties of a cryptographic hash function is that it is computationally infeasible to "predict" what the output hash value is given a specific input data. Besides, there is no statistical correlation between changes in the input data and changes in the resulting hash value.

In Bitcoin, for a system to accept a Bitcoin block as genuine, its block hash value has to be lower than the currently set target difficulty value. Changes, however minor, in the content of a block will result in a different block hash value; it is impossible for one to predict how the block hash value will change nor to craft the changes in the block's content such that the resulting block hash value is still lower than the target difficulty value.

In more blockchain systems, each new block is cryptographically chained to the previous block by including the previous block's hash value in the new block, as shown in Figure 16.10. As such, modifying the content of a previous block will result in a change in the previous block's hash value, thereby affecting the new block's hash value. In other words, modification on the content of block n will be reflected from block $(n + 1)$ to the latest block in the blockchain.

The private key is a unique key that only the owner can see. It is used by the sender of bitcoins to prove ownership of bitcoins. The private key(s) is(are) used to generate the public key. We use a private key like a password and anyone can verify that we know a "secret" (i.e., the private key) using our public key. The public key is visible to everyone. It is used by anyone in the community to verify one's ownership of bitcoins.

Blockchain

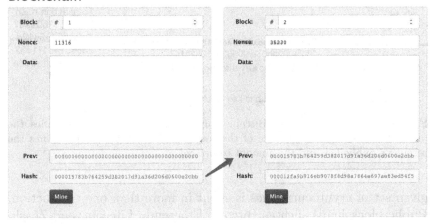

Figure 16.10: The hash of the previous block (Block #1) is stored in Block #2. *Source*: Brownworth (n.d).

The main usage of private and public keys is in the computation of digital signatures. What digital signature empowered is for the person with the private key to produce the correct signature. This correct digital signature will be verifiable with the corresponding public key, which then offers rights to perform bitcoin transactions.

16.4.1 *Transaction Hash*

In the blockchain, every transaction has its hash value. The hash value is computed using transaction data as input to the hash function. Depending on the specific blockchain system, the different hash functions will be used, and the input to the hash function also differs across designs. The key is that the transaction hash value serves as a unique identifier (analogous to the serial number on our physical bank notes) for a transaction. As with all payment systems, before a transaction is executed, its validity must be confirmed.

Validity checks include, among others, the verification that the sender has ownership of the amount to be spent, verification that the total value of the outputs must be smaller than or equal to the total value of inputs, verification that this transaction is not a double-spending transaction. A double spend is an attack where the

Field		Description
vin_sz		1
vout_sz		1
in	prev_out	7b844fe6a2ce9b1c7ea2f02bfb802a095ad3 352a092ac83aef0562ee5952b1d7
	scriptSig	Bob's public key and digital signature
out	value	5 btc
	scriptPubKey	Charlie's public key hash

Transaction hash:
317b9591b0a9d7
4afacd5735812d
236681e5111982
d2d57be21a598a
d1cba628

Figure 16.11: Example of Bitcoin transaction. The "in" field specifies the public key and digital signature of the sender and the "out" field specifies the receiver's public key and the amount to be sent.

given set of cryptocurrencies is spent in more than one transaction. Public blockchains such as Bitcoin, Ethereum, Litecoin, and Zcash, prevent double-spending by having a distributed ledger recording all transactions that have ever taken place and requiring at least 51% of the validating nodes to check if a given transaction has taken place sometime in the past.

Due to the property of the hash function that outputs transaction hash value that resembles a unique serial number, validating nodes simply check by performing a lookup of transaction hash values. An example of a Bitcoin transaction is shown in Figure 16.11.

16.4.2   *A Digital Chain of Blocks*

A blockchain is essentially a chain of blocks. It is a digital ledger, and the "digital chain" is enabled by the use of a hash function. Essentially, a hash value is calculated based on data in the block's header (we call this value the "block header's hash"). Such a notion is true for most major blockchain applications. Without loss of generality, we use Bitcoin as an illustration. Figure 16.12 shows the structure of a Bitcoin block. In the ith block, denoted as B_i, there is a list of bitcoin transactions.

To integrity-protect the block, it is enough to integrity-protect the block header. This is because the list of transactions is collectively "summarized" using a root hash stored in the block header; modifying any of the transactions will change the root hash, thus changes the data in the block header, causing a change in the resulting block header's hash.

Block i (B$_i$)

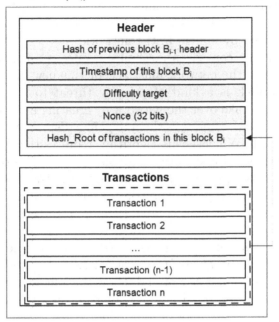

Figure 16.12: Bitcoin block structure.

Bitcoin does not employ a straightforward calculation of the block header's hash. Instead, the PoW consensus algorithm in Bitcoin requires miners to vary the 32-bit nonce and iteratively calculate the block header hash until the resulting hash value satisfies the difficulty target (see Figure 16.13). A valid Bitcoin block is where the block header's hash value is smaller than a fixed target value specified in the field "difficulty target".

As a secure hash function outputs a hash value that is pseudo-random and it is infeasible to predict which nonce would result in a valid block header hash, the PoW consensus algorithm is essentially a lottery system among miners. Each miner can only iteratively vary the nonce, re-compute the block header hash, and check if the block header hash is valid. This effort constitutes PoW. Each miner has an equal probability of finding a nonce that gives a valid block header hash. When a valid block header hash is found, the miner will broadcast the block in its entirety (referred to as "the solution") to the network. Other miners in the network will accept this block as valid by checking that after hashing the solution, the resulting block header hash satisfies (i.e., is smaller than) the difficulty target.

The value of the difficulty or target is part of the consensus rule globally accepted by miners in the network.

As part of the consensus mechanism, once a miner submits his PoW, the rest of the network must accept this new block.

A Sybil attack is a kind of security threat to a system where one user tries to take over the network by creating a large number of anonymous identities and uses them to gain a disproportionately high influence. The PoW consensus algorithm requires that the ability of a node (or a group of nodes) to create blocks must be proportional to the total processing power. This renders Sybil attacks economically impractical because a party who tries to launch a Sybil attack needs to have an extremely high computing power to keep creating new blocks.

Making changes to previous blocks on the Bitcoin network is also incredibly difficult because of the same logic. Under the PoW consensus algorithm, all nodes accept and follow the longest chain and work on the next block on top of it. If a node changes a previous block, it will have to create new blocks to append to the changed block at a faster speed than the rest of the world so that its chain will outgrow the original one. One will need at least 51% of the computing power of the whole network to achieve this goal.

The data structure of blockchain is a linked list (hence the term "chain") of data blocks (see Figure 16.14). Each block contains the hash value of the previous block, and this serves as a link to the previous block and establishes the order throughout the chain of blocks.

16.4.3   *Proof of Ownership of Wallets*

Recall the relationship between a user's private key, public key, and wallet address. Essentially, the order of generation is as follows:

$$\text{Private key} \rightarrow \text{Public key} \rightarrow \text{Wallet address}$$

For security purposes, private keys are truly random strings. From the private key, a user's public key is calculated using a one-way (irreversible) and pre-determined manner. The public key is then hashed and encoded to form the user's wallet address. As such, the

Block i (B$_i$)

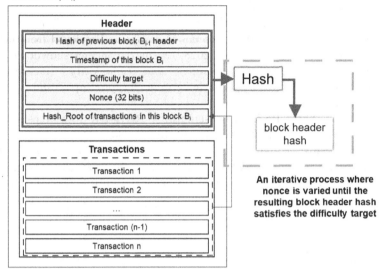

Figure 16.13: Bitcoin calculation of block header hash (PoW).

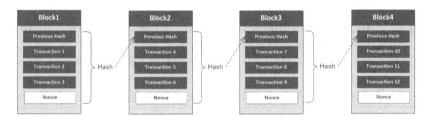

Figure 16.14: Blockchain data structure.

conclusion that "Bob has ownership to the wallet containing the 5 btc if his digital signature on the transaction can be verified using his public key" is based on the fact that Bob can only provide the (correct) public key if and only if he has the private key to create the digital signature at the first place.

Bibliography

Antonopoulos, A. M. (2017). *Mastering Bitcoin: Programming the Open Blockchain.* 2nd edition. O'Reilly Media, Sebastopol, CA.

Brownworth, A. (n.d). Blockchain demo. Retrieved from https://anders
brownworth.com/blockchain/blockchain.

Greenspan, G. (2018). *Smart Contract Showdown: Hyperledger Fabric vs
MultiChain vs Ethereum vs Corda*. MultiChain. Retrieved from https:
//www.multichain.com/blog/2018/12/smart-contract-showdown/.

Henderson, T. (2013). *Cryptography and Complexity*. Hackthology.
Retrieved from https://hackthology.com/cryptography-and-complex
ity.html.

Kaliski, B. S. (2001). RSA digital signatures. Retrieved from https://www
.drdobbs.com/rsa-digital-signatures/184404605.

Lo, S. W., Wang, Y., and Lee, D. K. C. (2021). *Blockchain and Smart
Contracts: Design Thinking and Programming for FinTech* (Singapore
University of Social Sciences-World Scientific Future Economy Series:
Volume 4). World Scientific, Singapore.

16.5   Sample Questions

Question 1

Which of the following statement is incorrect?

(a) Cryptographic techniques are secure because they are computationally infeasible to break at present
(b) A user's wallet address is not related in any way to the user's private key
(c) Hash function is used to obtain transaction and block hash values

Question 2

The six-block confirmation states that:

(a) User should wait for six additional blocks appended to the block where his/her transaction is in before regarding the transaction as confirmed
(b) Before validating a transaction, the miner should search in at least six other blocks to ensure that the transaction is not a double-spending transaction
(c) A successful miner who created a new block and obtained the block reward should wait for at least six blocks before he/she can spend the block reward

Question 3

Which of the following statement is not true about the hash function?

(a) The hash function is always deterministic
(b) The hash function is an encryption function
(c) The hash function is used to output a unique serial number for a transaction

Question 4

Which of the following statement is true about digital signatures?

(a) The proof of possession of bitcoin requires the signature and public key
(b) A digital signature is produced using the signer's public key
(c) The digital signature used in Bitcoin is the RSA digital signature

Question 5

Which of the following is not a property of the hash function?

(a) The same input always produces the same output
(b) A small change in the input will result in a drastic change in the output
(c) Given an input, it produces a truly-random hash value

Solutions

Question 1

Solution: Option **b** is correct.

A user's wallet address is generated in the following order: Private key → Public key → Wallet address. Thus, the wallet address is dependent on the user's private key but they cannot be linked.

Question 2

Solution: Option **a** is correct.

Question 3

Solution: Option **b** is correct.

The hash function is not an encryption function because it takes variable-length input and outputs a fixed-length hash value. This means that information is lost and it is unrecoverable (therefore, there is no "decryption" possible). Hash function serves the primary objective to integrity-protect data.

Question 4

Solution: Option **a** is correct.

Option b is incorrect because a digital signature is produced using the signer's private key. Option c is incorrect because Bitcoin uses an Elliptic Curve-based digital signature.

Question 5

Solution: Option **c** is correct.

A hash function produces only a pseudorandom hash value. It removes the statistical correlation between the input and the output, but the process is performed in a deterministic way.

Chapter 17

Consensus

Consensus is a key concept in Blockchain and Distributed Ledger Technologies. Blockchain networks are distributed in a manner whereby participants do not necessarily know or trust each other. This makes agreeing on the state of the network challenging, particularly when there may be bad actors. Consensus protocols are designed to achieve agreements within the network and to ensure that participants act in the best interest of the network.

17.1   Learning Objectives

- Discuss the characteristics of a distributed system and relate them to blockchain.
- Describe the fundamental properties of a consensus algorithm.
- Appraise the design of consensus algorithms for public and private blockchain.

17.2   Main Takeaways

17.2.1   *Main Points*

- A blockchain resembles a distributed system, which is characterized by properties such as concurrency, lack of a global clock, independent failure of components, and synchronous/asynchronous message-passing.

- Consensus is an important concept for blockchain networks and distributed ledgers as there is no central authority to provide the source of truth.
- A distributed consensus algorithm should possess or clearly define its performance in terms of synchrony, liveness and safety, and resiliency.
- The performance of consensus algorithms for public and private blockchain differ significantly from each other due to different permission and user authentication models.

17.2.2 *Main Terms*

- **Distributed system:** A distributed system involves a set of distinct nodes/computers passing messages to one another (via communication link) and coordinating to accomplish a common objective.
- **Distributed consensus algorithm:** An algorithm that allows members of a network to conclude the state of the network at a point in time by following some set of rules programmed in the protocol.

17.3 Distributed System and Characteristics

A distributed system involves a set of distinct processes (e.g., nodes/computers) passing messages to one another (via communication link) and coordinating to accomplish a common objective (i.e., solving a computational problem). The earliest form of a distributed system is the local area network, where a group of networked nodes shares a common goal for their work and each node essentially has its processor and memory. There are four properties of a distributed system (Kasireddym, 2018).

- *Concurrency*: The processes in the system operate concurrently, meaning multiple events occur simultaneously and independently. Concurrency control is a difficult problem in a distributed system because of latency (or lags).

- *Lack of global clock*: For a distributed system to work, we need a way to determine the order of events. However, in a set of computers operating concurrently, it is sometimes impossible to say which one of two events occurred first, as computers are spatially separated. In other words, there is no single global clock that determines the sequence of events happening across all computers in the network.
- *Independent failure of components*: Components in a distributed system may run into faults. There are many different types of fault, including crash, lost or incomplete messages, or even malicious messages. A crash fault is where nodes stop working because of a power outage or hardware fault, whereas a Byzantine fault is where nodes behave in an "adversarial context".
- *Message-passing*: Computers in a distributed system communicate and coordinate by "message-passing" between one or more other computers. There are two types of message-passing environments: synchronous and asynchronous. Synchronous message-passing means that the message will be delivered within some fixed, known amount of time, whereas asynchronous message-passing means that message may be delayed infinitely, may be duplicated, or delivered out of order.

To this end, we define the following:

- *Simple fault tolerance*: Assumes all nodes of the system either follow the protocol exactly or crash.
- *Byzantine fault tolerance*: Assumes nodes can crash or may choose to be malicious.

A blockchain is essentially a distributed system, and a distributed system is a network of computers working on the same problem concurrently and independently. In this system, there is no global clock that tells the order of events. Some of the computers may be honest, some of them may crash, and some are Byzantine nodes. There is also a lack of a reliable communication link and messages can be dropped or delayed indefinitely. The challenge lies in letting all nodes achieve consensus under this circumstance.

17.4   Motivations for Consensus

The research on consensus algorithms has been ongoing for many years. The earliest research is on the Practical Byzantine Fault Tolerance (PBFT) consensus, which serves as the foundation for the consensus algorithms of many existing permissioned blockchains. Then, there were the leader-based consensus mechanisms such as Paxos and RAFT (Seibold and Samman, 2016). Bitcoin's Proof-of-Work (PoW) consensus is a breakthrough in this area because it is the first consensus algorithm that scales gracefully as the number of nodes increases, and it allows the mutually distrusting group of nodes to agree to a common truth.

For a group to achieve consensus among its members, they should arrive together at a decision and agree to support it. The decision should be an acceptable resolution, a decision that the group can support even if it is not the preferred option of some individuals in the group. Government elections are one example of a group coming together to make such a decision.

In a distributed network of computers working together to maintain a database or ledger, consensus protocols (or algorithms) state the rules of the network and are followed by the computers in the network. Since blockchain networks are distributed in nature and participants do not necessarily know or trust each other, a consensus algorithm is a mechanism that allows the majority of these mutually distrusting nodes to agree on a common state of the blockchain. After transactions are formulated and broadcasted, validating nodes will verify and validate the transaction following the consensus algorithm.

Consensus is an important concept for blockchain networks and distributed ledgers as there is no central authority to provide the source of truth. The network members have to conclude the state of the network by following some set of rules. These rules are the consensus algorithms programmed into the blockchain protocol.

A public blockchain is susceptible to Byzantine faults and Sybil attacks. As mentioned earlier, a Sybil attack is a type of attack seen in peer-to-peer networks in which a node in the network operates multiple identities simultaneously and undermines the authority or power in reputation systems. A Byzantine fault is a type of fault that assumes that nodes can fail or be malicious. It includes data corruption faults or node collusion. Thus, the consensus algorithm must be fault-tolerant.

A good consensus protocol needs to maintain a consistent ledger. It should keep the network up-to-date, fix errors, and ignore malicious nodes. A blockchain network also needs to ensure that the transactions recorded on the ledger are valid. For consensus algorithms to achieve the objective of maintaining a valid blockchain ledger, it needs to be able to do the following:

- If bad actors try to broadcast invalid transactions, they should be ignored by the good nodes and not included in the ledger.
- When bad nodes try to mine a block with invalid or fraudulent transactions, the majority of the network should agree not to continue extending the blockchain on top of it.

In Figure 17.1, suppose there are four nodes on the network and the correct action is to "Proceed". Node B has the wrong information "Wait" and Node D is maliciously trying to get the nodes to "Cancel". A good consensus algorithm needs to update Node B such that it chooses "Proceed" and successfully ignores Node D.

To achieve this, the blockchain nodes need to:

- Be able to keep track of historical transactions
- Authenticate and validate current broadcasted transactions
- Reach agreement on what goes in the ledger with the other nodes
- Receive appropriate incentives to "do the right thing".

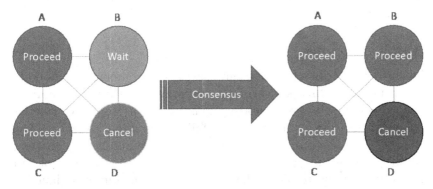

Figure 17.1: Example of a good consensus algorithm.

17.5   Major Consensus Algorithms

17.5.1   *The FLP Impossibility Theorem*

In their 1985 paper on "Impossibility of Distributed Consensus with One Faulty Process", researchers Fischer, Lynch, and Paterson (aka FLP) show how even a single faulty process makes it impossible to reach consensus among deterministic asynchronous processes (Fischer *et al.*, 1985). If we cannot assume a maximum message delivery time in an asynchronous environment, then achieving termination is much harder, if not impossible.

One method to overcome this is to use timeouts. In other words, if no progress is being made in the last time epoch, the consensus steps will start all over again when the timeout is reached. This relaxes the asynchronous requirement.

17.5.2   *Fundamental Properties*

Although PoW has probably earned the most fame from Bitcoin's success, it is hardly the only or the best consensus protocol (strictly speaking, it is not a distributed consensus protocol in the traditional sense) existence. Instead, it is designed to address a very special use case under a unique set of operating conditions that may not be applicable beyond the Bitcoin case.

For more than a decade, distributed consensus protocols have been an area of research in the realm of distributed systems. The only difference is that most of the previous research has never made it to a global scale like Bitcoin and operates within a highly asynchronous environment like the Internet.

17.5.2.1   *Synchrony*

Synchrony describes the message-passing environment of the nodes in the network. In a synchronous environment, there is a limit to message latency between nodes. This allows us to know whether messages are lost or nodes have failed based on timeouts. One example is video conferencing. In such an environment, consensus can be guaranteed to be reached as long as the number of failures is within the resilience level that the algorithm is designed to tolerate.

On the other hand, in an asynchronous environment, there is no fixed upper limit for message latency. One example is the Internet. In this case, even a single failure in the nodes makes it impossible to reach a consensus. This is known as the FLP impossibility result. However, by making assumptions like partial synchrony, it is still possible to find a solution to reach a consensus.

17.5.2.2 *Safety and Liveness*

All consensus protocols are designed to guarantee either one or both of these properties under different conditions. Safety relates to the consensus algorithm's ability to prevent the overwriting or the corruption of a previously validated state. In other words, nothing bad will happen. It is the fundamental requirement to ensure consistency in the system.

Liveness relates to the consensus algorithm's ability to guarantee that the message sent from the sending node will eventually reach the destination node. In other words, something good will eventually happen. It means that the blockchain can remain useful by continuing to validate and append new blocks.

Most consensus algorithms require synchrony for safety and liveness to be achieved. There is a trade-off between synchrony and liveness. The FLP impossibility means that any consensus algorithm can only have either liveness or safety, in addition to being fault-tolerant.

17.5.2.3 *Resiliency*

Resiliency is based on the maximum number of malicious nodes the consensus algorithm is designed to tolerate to guarantee safety and liveness under different fault conditions. For example, a resilience level of 33% means that at least two-thirds of the nodes must be honest.

17.5.3 *Consensus for Public Blockchain*

We first introduce consensus for public (and trustless) blockchains such as Bitcoin or Ethereum where none of the nodes is considered trusted. In such networks, appropriate incentives like mining rewards are needed to ensure that the network members behave correctly.

The incentives must be aligned to encourage the appropriate actions. Some general guidelines to consider:

- Doing the right thing gets rewarded.
- Rewards from continuing to do the right thing should outweigh the potential net benefits from doing bad things—the costs of doing bad things should be high.

In the following, we review three major consensus algorithms for public blockchain, namely proof-of-work, proof-of-stake, and delegate proof-of-stake, each having its pros and cons. In this recent article, Vitalik presented an analysis of the security of PoW and PoS (Vitalik, 2020).

17.5.3.1   *PoW*

In Bitcoin, the consensus algorithm used is PoW or Bitcoin mining. Nodes on the Bitcoin blockchain can choose which transactions to include into the block that they are mining or validating. They can also choose to ignore transactions that are broadcasted to the network. When a block is mined, the majority of nodes on the network agree on the state if they choose to mine on top of the previous block. Consensus is thus achieved when a block is accepted into the blockchain.

The PoW consensus algorithm was first introduced in Satoshi Nakamoto's Bitcoin Whitepaper. With the PoW algorithm, weak-form synchrony is achieved by controlling block production frequency to an average of 10 minutes and setting fixed boundaries for its timestamp. It is based on game theory and the scarcity of computational resources. Miners compete to solve cryptographic problems and append a block to the chain. The miner that contributes computing power and solves the problem will be rewarded by the network's digital currency. PoW is competitive as only one node can win the mining reward for each block. The steps in PoW are as followed and are described in Figure 17.2.

1. Receive new broadcasted transactions and check them against the ledger to ensure that they are legitimate (sender have signed the transaction and sender address have enough balance).
2. Discard the transaction if it is found to be invalid.

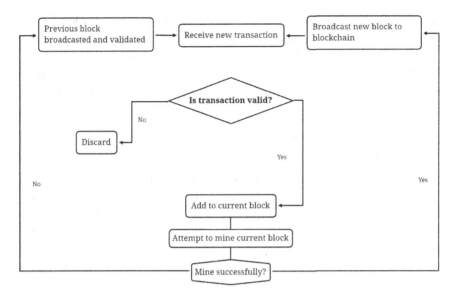

Figure 17.2: Steps in PoW consensus.

3. Package legitimate transactions into the current block.
4. Attempt to mine the block.
5. If it succeeds, broadcast the mined block to the network. Then start again from Step 1 and mine on top of the block you just created.
6. If not successful (the next block is broadcasted before yours), check if the new block broadcasted is valid, go back to step 1 and mine on top of the new block.

The act of mining is to solve a cryptographic puzzle that requires computation power. Transactions are not immediate because time is needed to solve the cryptographic puzzle. Mining is also not environmental-friendly as it requires a lot of energy.

The more computation power one has, the more permutations one can try per second and the higher the chance of mining the block. Competition in mining over the years has caused miners to amass large amounts of powerful hardware to increase their chances. This tends to favor the nodes with more resources as one miner using one hundred dollars of resources to mine will have higher chances than a hundred miners with one dollar each. This is also known as economies of scale. However, there are concerns that Bitcoin is not exactly decentralized due to large companies dominating the mining process. These large companies have the power to develop chips setting up mining pools that increase the chances of profitability through cumulative effort.

17.5.3.2 *Proof-of-Stake (PoS)*

The PoS algorithm is also commonly used in public blockchains. One needs to have a stake (or hold some coins) in the system to participate in PoS. Instead of miners mining new blocks, validators mint or forge new blocks. The PoS algorithm identifies a set of validating nodes that owns the most stakes that will validate transactions and forge new blocks. In its simplest form, if you own 5% of the total stake, your chance of mining the next block is also 5%.

PoS is more energy-efficient than PoW as not much computation (to solve a cryptography puzzle) is needed. PoS is also competitive; only one staker can mine each block.

Unlike PoW, PoS is not susceptible to economies of scale. The chances of mining the block are the same for someone with a hundred dollars' worth of coins and a hundred persons with a dollar each. Attacking a PoS is also more costly than PoW as you will lose your stake if the network detects that you are malicious. In PoW, you do not lose your coins or your mining hardware if you launch a malicious attack. An attacker in PoS needs to have enough stakes to launch an attack, which would be economically unfeasible since doing so would lead to a devaluation of the tokens (Lee and Low, 2018).

However, the main issue of PoS is the "nothing at stake" problem. In the event of a fork (where two miners create the next block at the same time), PoW miners have to choose the block they wish to mine on top of as their computing resources are limited. For PoS validators, there is nothing to lose to stake on both forks. When a fork occurs, the staker will have coins on both forks, allowing it to stake on both sides. This leads to no resolution on what is the correct chain (see Figure 17.3).

The industry generally looks to PoS as the most probable solution to the resource-hungry PoW, but the nothing-at-stake problem has to be resolved. There are various attempts to solve this, either by

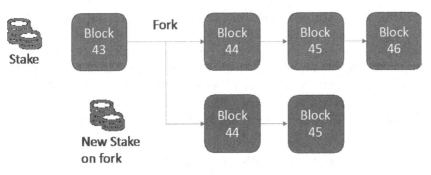

Figure 17.3: The nothing at stake problem in PoS.

using economic incentives or by punishments, such as the proposals in Ethereum 2.0.

Both PoW and PoS are used in public blockchain networks to ensure participants act in the best interest of the network by working to earn rewards. There are various iterations of these consensus protocols in the industry addressing different use cases. Permissioned blockchain also utilizes a different class of consensus protocols as their participants are generally trusted (unlike public blockchains).

17.5.3.3 *Delegated PoS*

Delegated proof-of-stake (DPoS) sounds very similar to the PoS framework, but the differences between them are distinct. Indeed, in principle, both conventions serve the objective of maintaining consensus across the entire system. However, DPoS adopts a different approach to achieve this objective. DPoS was first adopted in Bitshares after a proposal by Dan Larimer. How it function was through a set of nodes called witnesses.

These witnesses have to add new blocks to the blockchain. The panel witnesses are selected based on two systems: the reputation system and real-time voting. They roster their duty by taking turns to add new blocks at time t-interval. For example, for every 30 minutes, the list of witnesses will reshuffle and they will take a turn after every 30 minutes to add in new blocks. However, there is a constraint. The number of active witnesses allowed at one time is limited. The limitation varies according to the network, which usually ranges between 20 and 100 witnesses.

Under the DPoS framework, anyone is capable of downloading a full node and assist in the validation of transactions. However, this does not mean that they are rewarded. Participants have to be among the top N number of witnesses to qualify for the chance to add blocks and receive rewards. N is determined by the number of active users for a particular network. The top N participants are selected democratically by the number of votes each witness receives.

17.5.4   *Consensus for Private Blockchain*

In a distributed system (or blockchain) where users are trusted to act honestly, the corresponding distributed consensus algorithm can be relaxed in terms of liveness. In other words, the system may never reach a consensus. However, given the discussion in a private blockchain setting, this is highly unlikely to happen.

17.5.4.1   *Paxos*

Paxos is one of the most famous distributed consensus algorithms suggested by Lamport. It is widely used in distributed systems such as Chubby and ZooKeeper. The entities in Paxos include the Proposer, Acceptor, and Learner.

- Proposer—leader/coordinator
- Acceptor—listen to requests from proposers and respond with values
- Learner—learn the final values that are decided upon

In Paxos, messages are assumed to be delivered without being corrupted. Essentially, the process proceeds as follows:

1. Proposer asks Acceptors to prepare to accept the proposal with version number n.
2. Acceptors respond with an acknowledgment and will not accept any proposals with a version number greater than n. Acceptors will also respond with the value v of the proposal with version number n.
3. Proposer, after receiving responses from a majority of Acceptors, broadcast to all to accept (n, v).
4. Acceptors accept the proposal, responds to Learners about accepting (n, v). Learners learn the decided (n, v).

5. If Proposer crashes, the system will elect a new Proposer after a timeout.

It is worth noting that authentication of the Proposer, Acceptor, and Learner is important, because only when nodes are authenticated and can be held accountable that it is possible to identify the erring nodes in the event of Byzantine failure.

17.5.4.2 *Raft*

A node in the Raft consensus can only be in one of the three states at any point in time: leader, follower, and candidate. The process of Raft implementation is described as follows:

1. All nodes are Followers in the initial stage.
2. To become the Leader to propose, a Follower must become a Candidate and launch a round of electoral votes.
3. If the node does not receive enough votes, the node becomes a Follower again.
4. If it receives a majority of the votes, the node becomes the Leader.
5. If the Leader encounters failures and finds that a new leader is elected after it recovers from failures, the original leader automatically goes back to the follower state.

To maintain its authority, an elected leader must continuously send a heartbeat packet to the other nodes in the cluster. If a follower does not receive the heartbeat packet during a given election timeout, the leader is considered to have crashed and the follower changes its status to the candidate and starts a leader election.

17.5.4.3 *Practical Byzantine Fault-Tolerant*

The PBFT consensus was proposed in 1999 by Castro and Liskov. In a PBFT consensus, all nodes are ordered sequentially with one node being the *leader* node and others considered as *backup* nodes.

For a given proposal, all the nodes in a system must communicate with each other and reach a consensus based on the majority principle.

Each PBFT consensus round is called a view. The leader node is changed during every view and can be replaced with a protocol called a view change if a certain amount of time has passed without the leader node broadcasting the request.

- The Leader orders messages and propagates them through a three-step reliable broadcast to the replicas (backup nodes).
- Replicas monitor Leader for safety as well as for liveness (using timeout).
- Replicas can propose for view change (to elect a new Leader if the current one is unavailable or is deemed as malicious).
- All messages must be authenticated using a digital signature.

It is straightforward to notice that in Paxos, Raft, and PBFT consensus, the complexity of message communication increases drastically when the number of nodes increases; this affects the liveness property. In other words, in contrast to the consensus algorithms for public blockchain, the consensus algorithms for private blockchain will not scale well when the number of untrusted users increases.

Bibliography

Alibaba Cloud (2019). From distributed consensus algorithms to the blockchain consensus mechanism. Community Blog. Retrieved from https://www.alibabacloud.com/blog/from-distributed-consensus-algorithms-to-the-blockchain-consensus-mechanism_595315.

Fischer, M., Lynch, N., and Paterson, M. (1985). Impossibility of distributed consensus with one faculty process. Retrieved from https://dl.acm.org/doi/10.1145/3149.214121.

Kasireddym, P. (2018). Let's take a crack at understanding distributed consensus. Retrieved from https://www.preethikasireddy.com/post/lets-take-a-crack-at-understanding-distributed-consensus.

Lee, D. K. C. and Low, L. (2018). *Inclusive Fintech: Blockchain, Cryptocurrency and ICO*. World Scientific, Singapore.

Seibold, S. and Samman, G. (2016). *Consensus: Immutable Agreement for the Internet of Value*. KPMG. Retrieved from https://assets.kpmg/content/dam/kpmg/pdf/2016/06/kpmg-blockchain-consensus-mechanism.pdf.

Vitalik, B. (2020). Why proof-of-stake. Retrieved from https://vitalik.ca/general/2020/11/06/pos2020.html.

17.6   Sample Questions

Question 1
In a Byzantine Fault,

(a) a server can appear as both failed and functioning to fault-detection systems
(b) it is easy to shut out a component from the network
(c) consensus can still be reached

Question 2
Which of the following statement is false about Bitcoin consensus?

(a) If the sum of input values is less than the sum of output values, the bitcoin transaction will not be verified.
(b) There is only one golden nonce that makes the block header hash value less than the difficulty target.
(c) It can scale well; as more miners join the network, the network becomes more secure.

Question 3
A good consensus protocol should:

(a) Maintain a consistent ledger
(b) Produce as many transactions per second as possible
(c) Wait for all nodes to respond before finalizing the decision

Question 4
Which of the following affects the safety property of a consensus algorithm?

(a) Nodes did not respond in time to a proposed decision.
(b) Messages did not arrive within a pre-determined time.
(c) Nodes did not agree on the same state of the ledger.

Question 5
Which of the following is not correct about Practical Byzantine Fault-Tolerant protocols?

(a) Prioritizes safety
(b) Prioritizes liveness
(c) All users are authenticated

Solutions

Question 1

Solution: Option **a** is correct.

This behavior is a form of malicious activity as it aims to fool the network.

Question 2

Solution: Option **b** is correct.

Question 3

Solution: Option **a** is correct.

User authentication affects the throughput (or the number of transactions per second) of a blockchain platform; if the users are trusted, then a consensus can be reached faster via voting. It is also not necessary to wait for all nodes to respond; having a majority of nodes that agreed or voted on one decision is enough.

Question 4

Solution: Option **c** is correct.

Question 5

Solution: Option **b** is correct.

PBFT assumes all users are authenticated. Hence, most of the time, it is unlikely that users will disagree on a correct state. However, it may take longer for users to reach a consensus due to the number of communication rounds needed.

Chapter 18

Cryptocurrencies, Wallet, and Token Economy

18.1 Digital Currencies, Cryptocurrencies, and Tokens

Digital currency is not a new concept. With the success of Bitcoin, a decentralized cryptocurrency made possible with cryptography and blockchain technology, bitcoin together with other cryptocurrencies is gaining much attention. This section introduces the fundamental concepts about the digital currency, cryptocurrency, and token.

18.1.1 *Learning Objectives*

- Recognize the differences between digital currencies, cryptocurrencies, and tokens.

18.1.2 *Main Takeaways*

18.1.2.1 *Main Points*

- Cryptocurrencies and tokens are a kind of digital currency.
- The prime example of a digital currency is the Central Bank Digital Currency (CBDC).
- Cryptocurrencies and tokens are created on different blockchain layers; each can be further divided into several categories, depending on either its characteristics or purpose of usage.

18.1.2.2   Main Terms

- **Digital currency:** Digital representation of value denominated in their own unit of account.
- **Cryptocurrency:** A subset of digital currency that is based on blockchain and cryptography.
- **Token:** Does not have intrinsic value but is linked to an underlying asset, which could be anything of value. A token is commonly a representation of a digital asset.
- **Tokenization:** a process that transforms assets or data element into a token.

18.1.3   Value Transfer

We can classify value transfer broadly into four categories (Lee and Low, 2018).

Centralized, Not Geographically Bounded: Examples are loyalty points from financial, telecom, or retail companies, air miles from airlines, Second Life's Linden Dollar, and World of Warcraft Gold, which is a closed system with transactions within specific entities. This class of currency also includes cross-border pre-paid phone cards and to some extent, cash value smart cards, pre-paid debit, and credit cards. These cards can be physical or virtual. Other examples are the Alipay RMB wallet that can be used in different countries and even for tax refunds with QR (Quick Response) code or facial recognition. It is more appropriate to think in terms of online (e-wallet) and offline (physical card wallet) digital payments. This is a case of online/offline value transfer and storage unconstrained by location. This class may not be dependent upon governance as in fiat currency and more importantly, it is not geographically bounded.

Centralized, Geographically Bounded: Digitalized national currency, local or community currencies e-Brixton Pound and Bristol Totnes Pound that is used in England, and eChiemgaue in Germany. The purpose is more specific and usually bounded by some social contracts or agreements, such as honoring them for the exchange of goods or limiting the supply of goods. The governance is centralized and the value transfer is localized.

Centralized, Cross-platform: Flooz and Beenz, which are open market systems and can be transacted with other entities. Note that the crypto debit or credit cards such as TenX are built upon a decentralized system of cryptocurrencies and tokens. Smart contracts allow for the exchange of value between different digital currencies and across the network. The governance structure is centralized on top of decentralized and sometimes with smart contracts. The value is transacted digitally across the platform and can be online or offline.

Fully Decentralized or Distributed Currency: This includes the cryptocurrencies such as bitcoin, ether, Qtum, Zcash, Litecoin, Dogecoin, and others. They can be transacted with any outside agents, and the governance and technology are both decentralized due to open-source software. There is usually no legal entity responsible for the activities, and therefore they fall outside traditional regulation.

18.1.4 *Digital Currency*

Sometimes, the terms digital currencies and cryptocurrencies are used interchangeably, but they are not the same. A digital currency has all the characteristics of physical form currency, but it exists only in digital form. Digital currencies feature innovations that will unbundle the functions served by money (Brunnermeier *et al.*, 2019). The prime example that best illustrates digital currency is the central bank digital currencies (CBDC) (Lee and Teo, 2020; Lee *et al.*, 2021).

The Bank of International Settlement defines CBDC as follows:

> CBDC is central bank-issued digital money denominated in the national unit of account, and it represents a liability of the central bank. If the CBDC is intended to be a digital equivalent of cash for use by end-users (households and businesses), it is referred to as a "general purpose" or "retail" CBDC. As such, it offers a new option to the general public for holding money. CBDC is different from cash, as it comes in a digital form, unlike physical coins and banknotes. CBDC is also different from existing forms of cashless payment instruments for consumers such as credit transfers, direct debits, card payments, and e-money, as it represents a direct claim on a central bank, rather than a liability of a private financial institution. (Boar and Wehrli, 2021)

While retail CBDC is for general use, wholesale CBDC is for financial institutions and intended for the settlement of interbank transfers. According to a recent survey by the Bank of International Settlement

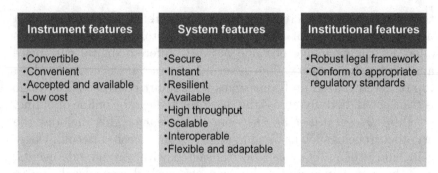

Figure 18.1: Features of digital currencies.

(BIS), the main motivations for CBDC include ensuring financial stability, monetary policy implementation, financial inclusion, and payments efficiency and safety. Financial inclusion and enhancing payments are key motivations for retail CBDC for nations located in emerging markets and developing economies. For wholesale CBDCs, the main motivation is to improve cross-border payments efficiency.

Figure 18.1 shows the core features of digital currencies (Bank for International Settlements, 2020).

With the COVID-19 pandemic in 2020, the push for digital currencies is stronger than before. The adoption of digital currency is now viewed by many as a necessity with the world's rapid digitalization. China is currently running trials of their CBDC, the Digital Currency Electronic Payment (DCEP) in four cities.[1] Bank of Thailand is testing and implementing Project Inthanon (Supadulya *et al.*, 2019) in the country, Project Inthanon-LionRock (Hong Kong Monetary Authority, 2020) with Hong Kong Monetary Authority, and mCBDC Bridge Project. This mCBDC is Phase 2 of the Project Inthanon-LionRock and involves the Hong Kong Monetary Authority (HKMA), together with the Bank of Thailand (BOT), the Central Bank of the United Arab Emirates (CBUAE), and the Digital Currency Institute of the People's Bank of China (PBC DCI), and strongly supported by Bank for International Settlement Innovation Hub Centre. It is a proof-of-concept (PoC) prototype to facilitate

---

[1]China's CBDC: Why this expert says the digital yuan is an "absolute necessity" (2020). Retrieved, August 12, 2020, from https://fortune.com/2020/07/30/china-digital-currency-yuan-cbdc/.

real-time cross-border foreign exchange payment-vs-payment transactions in a multi-jurisdictional context and on a 24/7 basis. While
CBDC can help reduce cross-border payment and settlement costs,
shorten the settlement time, ensure transaction security, enable
financial inclusion, domestic payments efficiency, and easy monitoring of the monetary policy implementation, some of the concerns are
the potential technological vulnerabilities and the impact on monetary and fiscal policy transmissions and exchange rates.

18.1.5 *Cryptocurrency*

Cryptocurrency is a peer-to-peer (P2P) programmable digital currency. It allows online payments to be sent directly from one
party to another without going through an intermediary. It is a
cryptographic-based digital currency and it is a subset of digital currency. Nowadays, it commonly refers to the digital currency that uses
cryptographic algorithms and blockchain technology. The definition
and categorization of cryptocurrencies vary from country to country.

The first centralized cryptocurrency, eCash (Chaum, 1983), was
introduced in the 1990s. Bitcoin is not the first digital currency that
uses cryptography, as the eCash used blind signatures[2] to protect
users' privacy years ago. Nonetheless, Bitcoin does come with innovation as it is the first decentralized cryptocurrency that relies on a
P2P network and solves the double-spending problem. The double-
spending problem refers to the issue that digital money can easily be
spent more than once. Bitcoin solves this problem by using cryptography, distributed consensus, and an immutable ledger. It provides
an alternative to those that have lost faith in a centralized monetary
system after the global financial crisis.

Since the launch of Bitcoin in 2008, thousands of other cryptocurrencies (or altcoins) have been invented for the past decade. *Altcoin*
is an abbreviation for Bitcoin Alternative, which refers to all other
cryptocurrencies that are not bitcoin. Some of them follow the source

---

[2]A blind signature disguises the content of a message before it is signed. As
a result, signer cannot view the message to be signed, but the resulting digital
signature can be verified by anyone. This is typically used in situation where the
signer and the message owner are different parties.

code of Bitcoin with small changes in the parameters such as maximum supply and block generation time or with added features such as privacy. Examples of altcoins are Litecoin and Zcash;[3] the latter added privacy features to Bitcoin and in the mining process, give some amount of block rewards to the developers instead of just the successful miners (Song, 2018).

Cryptocurrency can be an enabling tool to reach out to the unbanked and underbanked if appropriately designed. Cryptocurrency can be an excellent conduit for payments and funds, as well as fractional ownership of assets. Business is being transformed by diminishing the role of middleman, whether it is smart accounting or smart contract. When combined with the Internet of Things (IoT) and other technologies, it can maximize digitalization with digitization and digital identity. With decentralization and democratization of technology, services, and governance, it can lower cost to a level that allows the underserved and the excluded to enjoy goods and services previously unavailable to them.

There are thousands of cryptocurrencies that are actively traded in the market. The following subsections introduce a few most common ones based on their characteristics:

18.1.5.1 *Cryptocurrencies for Payment*

Bitcoin (BTC) was the very first and currently the most valuable digital currency. It relies on the collective effort of the entire network of nodes to confirm, verify, and record currency transactions. This is achieved via the clever use of cryptographic techniques in the proof-of-work (PoW) consensus mechanism, commonly known as "mining" and forming an immutable distributed ledger. The total number of bitcoin issuance is capped at 21 million, which is expected to be fully mined in 2140. The current circulation of bitcoin relative to the limit is more than 80% (as of August 31, 2020).

Ether (ETH) is the native currency of the Ethereum blockchain. Ethereum is a global decentralized application platform used for the development and operation of smart contracts. It is the

---

[3]CoinTelegraph. (n.d.). Altcoin News. Retrieved, February 27, 2021, from https://cointelegraph.com/tags/altcoin.

second-largest cryptocurrency (following bitcoin) in terms of market capitalization.

Litecoin (LTC) is an altcoin created to improve upon Bitcoin's shortcomings. Like Bitcoin, Litecoin is decentralized and relies on a P2P network. However, it produces blocks four times quicker than Bitcoin and has four times as large the supply of coins. This allows for faster transaction confirmation time using a different mining algorithm called Scrypt[4] in its PoW consensus mechanism.

Ripple (XRP) is an alternative platform to Ethereum. It does not operate on a blockchain network and it is based on a permissioned network (Bank for International Settlement, 2018). The usage of the Ripple platform is limited to a selected number of network nodes. XRP (Ripple Coin) is the base currency for Ripple network circulation. Anyone can create a Ripple account and transfer money to any currency through the Ripple payment network, including USD, EUR, RMB, JPY, and bitcoin. The transaction confirmation is completed in a few seconds and the transaction cost is almost zero. The maximum circulation of Ripple coins is 100 billion.

18.1.5.2 *Stablecoin*

Stablecoins are cryptocurrencies attempting to have minimal volatility in price. There are many studies such as that of Mita *et al.* (2019), Moin *et al.* (2019), and Moin *et al.* (2020) on how to reduce volatility in the price of stablecoins. For instance, a stablecoin can be backed by fiat currencies or commodities that have purchasing power. An example of such stablecoins is the *Gemini* dollar. Another method is to be backed by other cryptocurrencies, such as the MakerDAO's token, *DAI* that is collateralized with ether. The third method is to be tied to the algorithm that automatically changes the supply of the coins to keep the price stable (Bank for International Settlements, 2018).

It is important to note that while stablecoins might be backed by fiat currencies, they are not claimable on the balance sheet of a

---

[4]Scrypt is a password-based key derivation function. In contrast to Bitcoin's PoW, scrypt requires a large amount of memory to run, making it ASIC-resistant. Other cryptocurrencies that also use scrypt is Dogecoin and Auroracoin.

central bank because they are not issued by central banks (Bank for International Settlements, 2019).

18.1.5.3 *Privacy Coin*

A privacy coin is defined as a cryptocurrency that conceals data about its users. The Bitcoin blockchain data is publicly viewable. Bitcoin does not conceal traces of the cryptocurrency and does not hide the amount in each transaction. Consequently, Bitcoin is perceived as only pseudonymous, although its users hide behind pseudorandom strings of addresses, the transactions performed by each address are publicly viewable. Studies have shown that with social engineering, manipulations, or with prior knowledge on spending behavior, it is possible to deduce the identity of the real user behind a specific address (Khalilov and Levi, 2018).

Privacy coins provide a truly anonymous and private means for peer-to-peer payments. In private cryptocurrencies, the amount of cryptocurrencies traded and held in the wallets are usually not publicly viewable. *Zcash*[5] (ZEC) is an example of a privacy coin. It was created in 2016 and is based on Bitcoin's code. Zcash allows users to disclose private information partially and only if they wish to disclose it. It does so by employing "zero-knowledge proofs", or "zk-SNARK", a cryptographic algorithm that allows users to conceal transaction information such as the origin, destination, and amount. Other examples of privacy coins are Dash,[6] Monero,[7] and Grin.[8] It is important to note that these privacy coins operate on a public blockchain; the data, while viewable, are scrambled using cryptographic techniques.

The provision of privacy for cryptocurrencies is an important enabler for private financial transactions for individuals and businesses. This, however, brings about regulatory concerns on the effectiveness of existing anti-money laundering (AML) obligations

---

[5]Zcash (n.d.). Retrieved, February 27, 2021, from https://z.cash.
[6]Dash (n.d.). Retrieved, February 27, 2021, from https://www.dash.org.
[7]Monero (n.d.). Retrieved, February 27, 2021, from https://www.getmonero.org.
[8]GRiN (n.d.). Retrieved, February 27, 2021, from https://grin.mw.

on privacy coins. Dana *et al.* (n.d.) discussed how existing AML regulations can sufficiently cover risks such as money laundering and terrorist financing.

18.1.6   *Cryptocurrency vs Digital Currency*

Some of the main differences between cryptocurrency and digital currency are as follows.

Centralized vs decentralized: Most cryptocurrencies operate on a public and permissionless blockchain, but a traditional digital currency is usually manged via a centralized third party.

Anonymity and transparency: Most cryptocurrencies running on public blockchain allow either partial (e.g., Bitcoin) or complete (e.g., Zcash) anonymity. Transactions in public blockchains are also transparent except for privacy coin implementations. On the other hand, most digital currencies require user identification before transacting. The ledger is also not transparent and is only viewable by the centralized authority.

Legal framework: Cryptocurrencies such as Bitcoin, Litecoin, and Zcash are currently without a clear legal framework to govern their usages. On the other hand, as digital currencies research is commonly led by central banks, most countries have or are working on, some legal framework for digital currencies, i.e., Directive 2009/110/EC in the European Union.[9]

Remarks: The characteristics of digital currency detailed above are not absolute. Some CBDCs such as China's DCEP allow partial anonymity and offline transactions. As such, it is crucial to watch the space as various CBDC projects go into the pilot phase.

It can be tempting to correlate cryptocurrency and digital currencies. However, some view both as competitors of each other, and a primal argument for cryptocurrencies adoption is that they can serve as "an instrument to hedge risk from digital alternatives or

---

[9]CoinTelegraph (2017, December 13). Digital currencies vs. cryptocurrencies, explained. Retrieved from https://cointelegraph.com/explained/digital-currencies-vs-cryptocurrencies-explained.

cash propped up by central banks". (Huang, 2020) Some CBDC programs also plan to run on cryptocurrency platforms. The Reserve Bank of Australia, for instance, considered running its CBDC program on Ethereum (Bank for International Settlement, 2020).

18.1.7   *Token*

A token (or crypto-token) is a randomized data string without value or meaning, which serves as an identifier containing relevant information. Besides representing assets to be used in transactions, tokens can also represent utility. For example, if the token is operating on a video-sharing blockchain, it can entitle the user to stream a specified number of hours of content.

Sometimes it can be difficult to determine whether a cryptocurrency is a cryptocurrency (coin) or a token, with some using these terms interchangeably. To put it simply, coins generally refer to any cryptocurrency with its blockchain, while tokens refer to cryptocurrencies that are built on top of an existing blockchain. Some regulators classify coins as payment tokens. While coding on a new blockchain is easy with open-source code and allows for lower reliance on other teams to make regular technical enhancements, many high market capitalization projects do not have their original blockchain. For example, OmiseGo[10] is built on Ethereum's blockchain to improve Ethereum's scalability. Other tokens built on the Ethereum blockchain include the infamous CryptoKitties[11] and DAI mentioned earlier. Other token platforms include Stellar, NEO, and EOS.

Tokenization is a process that transforms various types of data into a token (or Cryp-token). It becomes feasible to use blockchain technology to make offline assets online—to tokenize them into a digital asset. For example, we can represent a cow with tokens to achieve the goal of tokenizing a cow such that a fraction of it can be traded on the market). Tokenization has immense potential to change

---

[10]OMG Network (n.d.). Retrieved, February 27, 2021, from https://omg.network.
[11]CryptoKitties (n.d.). Retrieved, February 27, 2021, from https://www.crypto kitties.co.

the way people invest in assets and allows for greater liquidity, faster and cheaper transactions, greater transparency, and accessibility.

There are many different ways to categorize tokens as shown in the studies by Kang *et al.* (2019), Lee (2019), Lo and Medda (2020), and Momtaz *et al.* (2019). According to the report by the Swiss Financial Market Regulatory Authority (FINMA, 2018), tokens can mainly be grouped into three types: payment token, function token (sometimes referred to as utility token), and asset token (sometimes referred to as asset-backed token or security token), with the possibilities of a hybrid between the three.

18.1.7.1 *Payment Token*

Payment tokens are synonymous with cryptocurrencies and have no other features or links to other development projects. These tokens may only develop the necessary functionality in some cases and become a means of payment over time. Examples of payment tokens include first-generation cryptocurrencies such as Doracoin and Litecoin, as well as tokens for payment and settlement, such as Dash.[12] Settlement tokens can be seen as a bridge between other digital assets and fiat currencies. When securities and fiat are represented using a token, the dependence on clearing and settlements house can be eliminated.[13]

18.1.7.2 *Function Token*

Function tokens, or utility tokens, are tokens designed to provide digital access to applications or services. Examples include Ripple and Stellar, with tokens for usage scenarios or potential usage scenarios. ERC-20 token, a type of Ethereum token, is the most commonly known utility token. On the Ethereum blockchain itself, gas

---

[12]Faridi, O. (2020, August 6). *Dash Teams Up with Taurus.io, a Mexican Digital Asset Exchange, to Launch a Cryptocurrency based Debit Card.* Crowdfund Insider. https://www.crowdfundinsider.com/2020/08/164961-dash-teams-up-with-taurus-io-a-mexican-digital-asset-exchange-to-launch-a-cryptocurrency-based-debit-card/.

[13]Sygnum (2020, August 5). Settlement tokens and their role in a tokenized ecosystem. https://www.insights.sygnum.com/post/settlement-tokens-and-their-role-in-a-tokenized-ecosystem.

is used to run smart contracts that execute business payment terms. Some decentralized applications (also known as DApps) might create ERC-20 tokens and use them for various functions in addition to payment. ERC-20 tokens are exclusively used on the Ethereum platform and follow certain standards so that they can be shared, exchanged, and transferred.

18.1.7.3 *Asset-backed Token*

Asset-backed or security tokens are investment contracts that represent legal ownership of a physical or digital asset, which must be verified within the blockchain.[14] Assets are such as participation in real entity gains, company shares or earnings interests, or the right to receive dividends or interest payments. In terms of their economic function, tokens are similar to stocks, bonds, or derivatives. For example, DGX tokens on DigixDAO have corresponding assets in the real world as the value of DGX is backed by actual gold stored in the company's reserve.

One way to determine whether tokens can be deemed as security tokens are through the Howey test.[15] In short, the Howey test has the following criteria for deeming transactions as investment contracts (U.S. Securities and Exchange Commission, 2019):

- It is an investment of money.
- The investment is in a common enterprise.
- There is an expectation of profit.
- The profit was to be generated from the work of the promoters or by a third party.

However, the Howey test is not a gold standard but a good reference because the DAO token failed to pass the test but was still deemed as securities by the Securities Exchange Commission (SEC). Several recent Commission enforcement actions involving "digital asset

---

[14]Mitra, R. (n.d.). Utility tokens vs security tokens: Learn the difference—ultimate guide. Retrieved, August 12, 2020, from https://blockgeeks.com/guides/utility-tokens-vs-security-tokens/.

[15]Singh, A. (2017). What is the howey test?—STARTUP BLOG. Retrieved from https://www.startupblog.com/blog/what-is-the-howey-test.

securities" illustrate the importance of interpreting what constitutes a security token. Regulations and interpretations are different across jurisdictions, and consultation with legally trained professionals is essential to avoid any regulation or law violations.

18.1.7.4 *Fungibility*

Fungibility refers to the ability of various kinds of goods or assets to be mutually substituted with each other. It is an important concept that distinguishes some tokens from cryptocurrencies. Fungible tokens mean that the holders of the tokens do not care about which specific type of token they own as they can be readily interchanged, as per common understanding of currencies. Most tokens are fungible as users and investors do not differentiate among them. For example, we say "a certain amount of bitcoins", but we do not distinguish one bitcoin from another bitcoin as they are viewed as the same. Ether on the Ethereum blockchain is also a fungible token (the ERC-20[16] standard defines features of a fungible token created on Ethereum that can be programmed into a smart contract).

Non-fungible tokens (NFTs) refer to tokens that have distinguished identities and cannot be substituted by others. They can be used in DApps such as crypto-collectibles or crypto-games, or be used as asset tokens, access tokens, access transfer tokens and identity tokens, and certificates. This feature allows the tokenization of assets to have a wider application. NFTs enable the tokenization of any asset at a low cost. They can also represent unique certificates, keys, passes, wills, and any type of access right, loyalty programs, medical data, and many more other applications. An example of an NFT is the CryptoKitties token, where each token is unique to each CryptoKitty. On the Ethereum blockchain, the ERC-721[17] standard is one of the standards for representing NFT.

---

[16]Ethereum. (2020, December 07). ERC-20 token standard. Retrieved from http s://ethereum.org/en/developers/docs/standards/tokens/erc-20/.

[17]Ethereum. (2021, January 15). ERC-721 non-fungible token standard. Retrieved from https://ethereum.org/en/developers/docs/standards/tokens/erc-721/.

18.2   Wallets

18.2.1   *Introduction*

A cryptocurrency wallet is a tool that can be used to interact with blockchains. It stores the user's private keys and allows the user to send and receive cryptocurrencies, track the balance, and manage the keys and addresses. These wallets do not physically (or electronically) store the cryptocurrencies, but merely function as a method for key storage. Although we specifically distinguish the differences between tokens and coins, the transactions involving tokens and coins will originate from the user's wallet. As of June 2020, there are over 50 million blockchain wallet users worldwide, and that number keeps increasing.[18]

18.2.1.1   *Learning Objectives*

- Recognize different wallet implementations and their features.

18.2.1.2   *Main Takeaways*

18.2.1.2.1   Main Points

- A wallet is a tool that users can use to track their balance, send/receive cryptocurrencies, and manage their private keys.
- Wallets' implementation depends on the respective service provider.
- The hardware wallet is a type of cold storage; cold storage for cryptocurrency exchanges uses general-purpose computing hardware.

18.2.1.2.2   Main Terms

- **Non-deterministic wallet:** A wallet that generates private keys that are random and independent from each other.

---

[18]Statista. (n.d.). Number of Blockchain wallet users worldwide from November 2011 to February 22, 2021 (in millions). Retrieved, August 26, 2020, from https://www.statista.com/statistics/647374/worldwide-blockchain-wallet-users/#:~:text=The%20number%20of%20Blockchain%20wallets,the%20end%20of%20June%202020.

- **Deterministic wallet:** A wallet that adopts a predefined standard algorithm and uses a master private key to derive a set of private keys.
- **Hot wallet:** A wallet that is connected to the internet. Hot wallets are easier to use but is more susceptible to cyberattacks compared to cold wallet.
- **Cold wallet:** A wallet that is not connected to the internet.
- **Cold storage:** A method commonly adopted by cryptocurrency exchanges to store bitcoins offline and that the private keys that can unlock the bitcoins are never online.

18.2.2 Types of Wallets

18.2.2.1 Non-deterministic vs Deterministic

A non-deterministic wallet generates independent private keys that are not related to each other. Each generation process will use a different random number. A deterministic wallet, on the other hand, generates keys that are derived from a single master secret key (also known as the seed). All the keys in this wallet can be regenerated as long as the master secret key is not lost (Antonopoulos and Wood, 2018).

In terms of usability, a non-deterministic wallet requires the user to maintain a long and growing list of private keys securely. The deterministic wallet eliminates this concern as a user can have multiple private keys and can always recover the private keys as long as the master secret key is kept secure. There are many reasons that a cryptocurrency user prefers to possess more than one private key, that is, addresses. It is known that although on public blockchains like Bitcoin and Ethereum, users are anonymous and identity is concealed behind a random wallet address, reusing the address reveals information about spending behavior. Given that all transactions on a public blockchain are publicly viewable, user privacy can be compromised when sufficient data about an address is obtained.

An example of a deterministic wallet is the hierarchical deterministic (HD) wallet defined by Bitcoin's BIP-32 standard. The HD wallet standard specifies how keys can be generated from a master secret key. To encode a private key for secure back-up and retrieval, an approach employing mnemonic words (BIP-39) is proposed. This approach encodes the master private key into a list of English (or other languages) words. Using the mnemonic words allows one to recover the encoded private key. As such, the common and recommended practice is to record the list of words on a piece of paper and store the paper away securely.

18.2.2.2   *Hot vs Cold Wallet*

Hot wallets refer to wallets that have a connection to the internet. The benefit of using a hot wallet is the ease of use as the internet connection allows the wallet to incorporate all the applications used in transactions, making it more convenient for the user when doing frequent trading. Unfortunately, the user's private information is generally stored on a relatively centralized internet server and is thus more prone to attack from hackers than cold wallets. A software wallet is a typical example of a hot wallet and its categories are summarized as follows.

- *Web wallet*: Web wallets allow users to interact with the blockchain via a website. This means that the user need not download any proprietary software to gain access to blockchain interaction. SoFi[19] is an example of a Web wallet.
- *Desktop wallet*: Desktop wallets are wallet applications that users can download to their devices and gain access. In general, desktop wallets are considered safer than Web wallets. Electrum[20] is an example of a desktop wallet.
- *Mobile wallet*: Mobile wallets are mobile applications that allow users to interact with the blockchain. Web wallets would generally have a mobile application to be considered a kind of mobile wallet. Coinomi[21] is an example of a mobile wallet.

---

[19]SoFi (n.d.). Retrieved, February 17, 2021, from https://www.sofi.com/invest/buy-cryptocurrency/.
[20]Electrum (n.d.). Retrieved, February 17, 2021, from https://electrum.org/#home.
[21]coinmi (n.d.). Retrieved, February 17, 2021, from https://www.coinomi.com/en/.

Cold wallets are the opposite of hot wallets. They are wallets that do not have any connection to the internet. The following are two common forms of cold wallets.

- *Hardware wallets* are physical, electronic devices that use random number generators (RNG) to generate public and private keys.
- *Paper wallets* are an offline mechanism. Users record their private keys on a printed piece of paper, usually in the form of QR codes. Paper wallet holders can access the keys for their transactions by scanning the QR code. While popular in the early years of bitcoins for their security away from the internet, paper wallets are no longer deemed useful with other available wallets due to increased risks associated with losing, damaging, and misreading the paper wallet as well as its inability to send partial funds per transaction.

18.2.2.3 *Single or Multiple Cryptocurrencies Wallet*

Most wallets have the features of storing more than one cryptocurrency. In the study by Hileman and Rauchs (2017), more than one-third of wallets can store more than one cryptocurrency, and about 20% of wallets can store more than three cryptocurrencies. The vast majority of wallets support bitcoin and the few most popular cryptocurrencies such as Ether, Litecoin, and dogecoin. To this end, note that many cryptocurrency wallets integrated services of cryptocurrency exchanges to facilitate trading.

18.2.2.4 *Multi-signature Wallet*

Multi-signature wallets are wallets that require two or more private keys to sign on a transaction before the transaction can be sent to the network for validation. Standard transactions require only a single signature created by the sender. However, an M-of-N transaction requires at least M out of the N specified senders to sign before the transaction can be sent. The primary objective of this design is for security purposes as detailed on this page.[22] A multi-signature wallet provides such functionality to create an M-of-N transaction.[23]

18.2.2.5 *Cold Storage*

Cold storage is a method commonly used by cryptocurrency exchanges to secure their cryptocurrencies reserve (Hileman and Rauchs, 2017). A cold storage system works by keeping private keys in devices that are completely cut-off from any form of connectivity, whether wired or wireless.

Cold storage commonly works using an HD wallet, where the exchange holds a master private key and the corresponding master public key. For reserves designated to cover anticipated withdrawals in a day, only the corresponding private keys are kept on the online server while the rest are secured in cold storage. The exchange then uses the master public key to create a watch-only wallet. Due to the properties of HD wallet, the master public key can be used to provide addresses for receiving cryptocurrencies. To cater for excess withdrawal, if needed, the watch-only wallet will create a transaction without the signature (because the private key is in cold storage). Then, the unsigned transaction will be transferred to the offline computer, where the wallet on the offline computer returns a signed transaction.[24]

18.2.3 *Challenges and Open Question*

One of the prominent features of blockchain is that it is immutable, meaning that it is next to impossible to alter or change it through any means. Therefore, should a hacker gain access to your private key, he/she could use the private key to create transactions from your wallet and send all your funds to another location, and

---

[22]Multisignature. (n.d.). Retrieved, February 17, 2021, from https://en.bitcoin.it/wiki/Multisignature.

[23]Alex. (n.d.). What is a multisignature (multisig) or shared wallet? Retrieved, February 15, 2021, from https://support.bitpay.com/hc/en-us/articles/3600326 18692-What-is-a-Multisignature-Multisig-or-Shared-Wallet-.

[24]Cold storage. (n.d.). Retrieved, February 11, 2021, from https://en.bitcoin.it/wiki/Cold_storage.

there is little that you can do to reverse that transaction. Therefore, while the blockchain itself may be very secure, its weakest link is the way the users safely keep the private key. The security of one's private key varies, depending on the type of wallet used. Often, users must juggle between their wallet's security and its ease of use.

The survey by Hileman and Rauchs (2017) showed that about 32% of surveyed wallets are "closed source", which means that the source code of the wallet is not available for inspection by the developers' community. A worrying trend is that all custodial wallets are closed source. Similarly, 11% of self-hosted wallets are closed source as well. The same concern applies to hardware wallets.[25]

Wallets are also subjected to compliance standards and regulations. Specifically, to own a wallet, most wallet services will require the user to perform a KYC verification. The Financial Action Task Force (FATF) seeks to also impose a traditional anti-money laundering (AML) framework onto virtual asset service providers (i.e., wallet services) (FATF, 2020).

18.3 Token Economy

18.3.1 *Token Economy Basics*

Token economy refers to an innovative decentralized network and system incentivized by the use of crypto-tokens. It has to do with the study of human and system behavior. Reward mechanisms reinforce the participants' behavior and are designed to encourage target behavior such as checking the correctness of transactions or ensuring a steady supply of tokens in the network. Penalizing mechanism reduces bad behavior so that people will pursue more good behavior and avoid undesirable behavior autonomously in a peer-to-peer decentralized system.

---

[25]Rapoza, K. (2020, December 28). After ledger hack, who can you trust for bitcoin storage? Retrieved from https://www.forbes.com/sites/kenrapoza/2021/12/28/after-ledger-hack-who-can-you-trust-for-bitcoin-storage/?sh=4e06d1953840.

It is important to specify the cap on total supply and new tokens issuance method in token economy design. Also, the signaling of targeted behavior needs to be specific, and the reward needs to be transparent and clearly defined. Clear and transparent rules of (i) what to do to earn a token; (ii) how many tokens are earned; or (iii) how to redeem rewards (or back-up reinforcers) also should be specified beyond doubt for a token economy to function as intended. In other words, it must be designed in such a way that the tokens are worth holding in the foreseeable future.

This section is a brief introduction to understand the token economy. More details about the token economy design and token economics will be covered in Level 2.

18.3.1.1 *Learning Objectives*

- Discuss the objectives of the token economy and appraise token economics design.

18.3.1.2 *Main Takeaways*

18.3.1.2.1 Main Points

- Token serves as an enabler for a sustainable economic model.
- Token economy refers to a mechanism that incentivizes certain actions to reward good behaviors and penalize bad behaviors.
- The three key elements of a successful token economy are the tokens, clearly defined target behaviors, and backup reinforcers.
- Features of a successful token economy include the following features: facilitating the ecosystem, high value, listed on exchanges, resistance to inflation, social and technical scalability, and growth potential.

18.3.1.2.2 Main Terms

- **Token economy:** A network that incentivizes good actions with rewards and penalizes bad behaviors.
- **Mechanism design:** An approach that defines the rules and corresponding actions to get strategic outcomes.

18.3.2 *The Role of Token*

A token can be a symbol of value or right. As the symbol of value, the token is the core of the entire economic model, while the token economy considers the creation and distribution of the value, the consumption, and the circulation. As a symbol of right, a token represents a kind of incentive, while the token economy considers what type of organization is suitable, how to create a compelling economics model to embrace partners, suppliers, or even competitors and contribute to the whole ecosystem, and what the changes to the governance model are (Lo *et al.*, 2021). Besides the roles of the tokens, the future trend of the token and token values is essential for sustainability.

As similarly echoed by Sunyaev *et al.* (2021) and Kim (2018), tokens are viewed as the enabler and drivers of sustainable development. Tokens act as moderators that smoothen the flow of operations and by ensuring that an economy in which the total supply is "regulated" by code creates sustainable societies from an economic and social perspective.

Tokens can be acquired via several means. As a user of the blockchain platform, one may acquire tokens by providing services (such as validating transactions and extending the blockchain), or by investors at the time of their issuance. This is commonly through an event called ICO (see Chapter 19). Tokens can also be purchased on the secondary market.[26] Nevertheless, the objective of the token issuer is to ensure that the token economy design is sound.

18.3.3 *Effective Token Economy*

For a token economy to be effective, three elements are necessary: tokens (to be used as reinforcers to exchange for other reinforcers), back-up reinforcers (that act as rewards), and specified target behaviors. In particular, a token is digital or virtual. It has no intrinsic

---

[26]Entremont, P. (2017). Token Economy 101, or why Blockchain-powered decentralized networks are important. Retrieved from https://pentremont.medium. com/token-economy-101-or-why-blockchain-powered-decentralized-networks-ar e-important-310de1cc8bac.

value, but it can be exchanged for other material reinforcers, services, or privileges (various forms of back-up reinforcers that act as rewards). So, the tokens earned when people exhibit good behaviors can positively encourage good behaviors.

Three key elements of a successful token economy:

- *Tokens*: to be used as reinforcers to exchange for other reinforcers
- *Target behaviors*: specified and clearly defined
- *Back-up reinforcers*: to act as rewards and possible forms include things, services, and privileges.

When designing a token economy, the functions of the tokens need to be examined thoroughly so that the ecosystem can function well and be sustainable. In a successful token economy, well-developed tokens have the following features:

- Facilitating the ecosystem
- High value
- Listed on exchanges
- Resistant to inflation
- Social and technical scalability
- Potential for growth

The term "token economics" stems from mechanism design. Mechanism design is an economics approach to design incentives (or economic mechanisms) to achieve a strategic outcome (where players act rationally). One key consideration for the token design is the creation or generation of the tokens, how the tokens are distributed to the network (is it competitive), and the utility of the token within the network. Bitcoin's mechanism design aims to get miners to participate in validating transactions, ensures the blockchain ledger is consistent, and prevents double-spending of coins. The incentive used were mining rewards of bitcoins, fueled by its limited supply. This, in turn, gives bitcoins an economic value and drives miner incentives.

To design a token economy and consensus, the key consideration is the desired outcome. The network participants' incentives must be aligned to drive toward that outcome, regardless of whether mining or other methods are used as an incentive. One should also be on the lookout for undesired outcomes. When Bitcoin was designed, it was not expected that there would be a consolidation of miners. When bitcoin prices, in reality, became extremely attractive, it led to a centralization of mining power and the exclusion of nodes with low computational power. It was also not expected that the intense competition in mining would lead to such a large use of electricity consumption for the network.

In some instances, the desired objectives may not be just the actions that happen within the blockchain network but also the token price, such as keeping the price stable. The token design can affect the price. For example, the tokens' scarcity or high utility can contribute to a potential increase in the prices. It is important to note that token prices are affected by factors external to the network as well since non-participants may trade the tokens and contribute to price volatility. If you intend for the token to power the utility within the blockchain network, you need to consider its price volatility in the external market. Bitcoin's volatile and speculative nature has prevented it from fulfilling its original intention as a payment token.

Bibliography

Ahmad, S., Nair, M., and Varghese, B. (2013). A survey on cryptocurrencies. In *The 4th International Conference on Advances in Computer Science*, AETACS, Citeseer, pp. 42–48.

Antonopoulos, A. M. and Wood, G. (2018). *Mastering Ethereum: Building Smart Contracts and Apps*. O'Reilly Media, Sebastopol, CA.

Bank for International Settlements (2018). Central bank digital currencies. Committee on Payments and Market Infrastructures. Retrieved from https://www.bis.org/cpmi/publ/d174.pdf.

Bank for International Settlements (2019). Investigating the impact of global stablecoins. G7 Working Group on Stablecoins. Retrieved from https://www.bis.org/cpmi/publ/d187.pdf.

Bank for International Settlements (2020). Central bank digital currencies: foundational principles and core features. Retrieved from https://www.bis.org/publ/othp33.pdf.

Boar, C. and Wehrli, A. (2021). Ready, steady, go? Bank for International Settlements Survey on CBDC.

Brunnermeier, M. K., James, H., and Landau, J. P. (2019). The digitalization of money (No. w26300). National Bureau of Economic Research.

Chaum, D. (1983). Blind signatures for untraceable payments. *Advances in Cryptology Proceedings*. 82(3): 199–203.

Dana, V. S., Joshua, L. B., and Nick, L. (2021). Anti-money laundering regulation of privacy-enabling cryptocurrencies. Retrieved from https://www.perkinscoie.com/images/content/2/3/v7/237411/Perkins-Coie-LLP-White-Paper-AML-Regulation-of-Privacy-enablin.pdf.

ETH: Ethereum Whitepaper. Retrieved from https://ethereum.org/en/whitepaper/.

FATF. (2020). Guidance for a Risk-Based Approach to Virtual Assets and Virtual Asset Service Providers. Retrieved from https://www.fatf-gafi.org/publications/fatfrecommendations/documents/guidance-rba-virtual-assets.html.

FINMA. (2018). Developments in FinTech. Retrieved from https://www.finma.ch/en/documentation/dossier/dossier-fintech/entwicklungen-im-bereich-fintech/#:~:text=As%20set%20out%20in%20its,of%20money%20or%20value%20transfer.

Hileman, G. and Rauchs, M. (2017). Global cryptocurrency benchmarking study. *Cambridge Centre for Alternative Finance*, 33, 33–113.

Hong Kong Monetary Authority (2020). The Outcomes and Findings of Project Inthanon-LionRock and the Next Steps. Retrieved from https://www.hkma.gov.hk/eng/news-and-media/press-releases/2020/01/20200122-4/.

Huang, R. (2020). Central Bank Digital Currencies are Not Cryptocurrencies. Retrieved from https://www.forbes.com/sites/rogerhuang/2020/10/13/central-bank-digital-currencies-are-not-cryptocurrencies/?sh=6054bb3b27a3.

Kang, S., Cho, K. and Park, K. (2019). On the Effectiveness of Multi-Token Economies. *IEEE International Conference on Blockchain and Cryptocurrency (ICBC)*, 180–184.

Khalilov, M. C. K. and Levi, A. (2018). A survey on anonymity and privacy in bitcoin-like digital cash systems. *IEEE Communications Surveys & Tutorials*, 20(3), 2543–2585.

Kim, J. (2018). Crypto Token Economy Design for Disruptive BM. Retrieved from http://site.ieee.org/bcsummitkorea-2018/files/2018/06/D1_SKT_Crypto-Token-Economy-Design-for-Disruptive-BM_Jongseung-Kim.pdf.

Lee, J. Y. (2019). A decentralized token economy: How blockchain and cryptocurrency can revolutionize business. *Business Horizons*, 62(6), 773–784.

Lee, D. K. C. and Low, L. (2018). Inclusive fintech: Blockchain, cryptocurrency and ICO. World Scientific.

Lee, D. K. C. and Teo, E. G. (2020). The new money: The utility of Cryptocurrencies and the need for a New Monetary Policy.

Lee, D. K. C., Yan, L. and Wang, Y. (2021). A global perspective on central bank digital currency. *China Economic Journal*, 1–16.

Lo, S. W., Wang, Y. and Lee, D. K. C. (2021). Blockchain and Smart Contracts: Design Thinking and Programming for Fintech (Vol. 4). World Scientific.

Lo, Y. C. and Medda, F. (2020). Assets on the blockchain: An empirical study of tokenomics. *Information Economics and Policy*, 53, 100881.

Mita, M., Ito, K., Ohsawa, S., and Tanaka, H. (2019). What is stablecoin?: A survey on price stabilization mechanisms for decentralized payment systems. In 2019 8th IEEE International Congress on Advanced Applied Informatics (IIAI-AAI), 60–66.

Moin, A., Sekniqi, K. and Sirer, E. G. (2020). SoK: A classification framework for stablecoin designs. In International Conference on Financial Cryptography and Data Security Springer, Cham, 174–197.

Moin, A., Sirer, E. G. and Sekniqi, K. (2019). A classification framework for stablecoin designs. arXiv preprint arXiv:1910.10098.

Momtaz, P. P., Rennertseder, K. and Schröder, H. (2019). Token Offerings: A Revolution in Corporate Finance? Working paper.

Nakamoto, S. (2008). Bitcoin whitepaper. Retrieved from: https://bitcoin.org/bitcoin.

Ripple: Todd, P. (2015). Ripple protocol consensus algorithm review. Ripple Labs Inc White Paper (May, 2015) https://raw.githubusercontent.com/petertodd/rippleconsensus-analysis-paper/master/paper.

Song, J., (2018). Why Hard Forks are Altcoins. Retrieved from https://jimmysong.medium.com/why-hard-forks-are-altcoins-e0d3836c900d.

Sunyaev, A., Kannengießer, N., Beck, R. *et al.* (2021). Token Economy. Business and Information Systems Engineering. Available at https://link.springer.com/article/10.1007/s12599-021-00684-1.

Supadulya, C., Tansanguan, K. and Sethaput, V. (2019). Project inthanon and the project DLT scripless bond. Retrieved from https://www.adb.org/publications/project-inthanon-and-project-dlt-scripless-bond.

U.S. Securities and Exchange Commission (2019). Framework for Investment Contract Analysis of Digital Assets. Retrieved from https://www.sec.gov/corpfin/framework-investment-contract-analysis-digital-assets.

18.4   Sample Questions

Please select the most appropriate response.

Question 1

Which of the following is not a feature of fungible tokens?

(a) Divisible

(b) Unique

(c) Interchangeable

Question 2

Which of the following statement is false about cryptocurrency wallets?

(a) Cryptocurrency wallets include: Software wallets, hardware wallets, and paper wallets

(b) Paper wallets are preferred to hardware wallets as they can send partial funds

(c) Mobile wallets are the same type of wallet as website wallets

Question 3

Which of the following is a stablecoin?

(a) Monero

(b) DASH

(c) DAI

Question 4

Which of the following statement is true about deterministic wallets?

(a) A deterministic wallet uses one master secret key to generate private keys for multiple users

(b) An example of a deterministic wallet is specified in BIP-39 standard

(c) A deterministic wallet uses one master secret key to derive private keys

Question 5

Which of the following statement is false about token economy?

(a) The design should take into account the number of users on the platform

(b) The design should consider future trend and value of the token

(c) The design should properly define the desired outcome for behavior governance

Solutions

Question 1

Solution: Option **b** is correct.

Fungible tokens are identical, interchangeable, and divisible.

Question 2

Solution: Option **b** is correct.

Paper wallets are unable to send partial funds. They are also less recommended than hardware wallets.

Question 3

Solution: Option **c** is correct.

DAI is a stablecoin that is collateralized with ether. Monero and DASH are privacy coins.

Question 4

Solution: Option **c** is correct.

A deterministic wallet belongs to only a single user; the master secret key should be kept secret and known by that user only. An example of a deterministic wallet is the HD wallet defined by Bitcoin's BIP-32 standard.

Question 5

Solution: Option **a** is correct.

The design does not need to consider the number of users but on the characteristics of the users and how to properly reward good behavior and penalize bad behavior.

Chapter 19

Trading, Market, and Investment

19.1 Trading and Market

While cryptocurrencies might give people the impression that they are a form of currency, many countries' regulators do not perceive cryptocurrency as a form of traditional currency. One of the reasons is that cryptocurrencies do not fulfill the criteria of a storage of value, as the value of a cryptocurrency may fluctuate quite significantly over short periods.

For example, one bitcoin was worth about $7,100 at the end of 2019 but reached a high of $19,600 at the end of 2018. Such fluctuations make paying for goods with cryptocurrencies a risky proposition. Many people treat cryptocurrency as a tradeable asset, much like a stock or an alternative investment. In the following section, we will go through the process of trading cryptocurrencies and the various terminologies one might encounter when doing so.

19.1.1 *Learning Objectives*

- Relate the process of trading cryptocurrencies and their various terminologies.
- Describe typical psychological behaviors when trading cryptocurrencies.

19.1.2 *Main Takeaways*

19.1.2.1 *Main Points*

- Like securities, cryptocurrencies can be traded on exchanges known as cryptocurrency exchanges.
- One can exchange fiat currency for cryptocurrencies at these exchanges, which could subsequently be used to trade for other cryptocurrencies.
- Derivative trading can be used to invest in cryptocurrency indirectly.
- As cryptocurrency is a relatively nascent investing option, traders must beware of market manipulators who could resort to malicious trading methods such as pumping and dumping to earn quick profits at the expense of the general investing public.

19.1.2.2 *Main Terms*

- **Market Orders:** Orders to buy/sell the cryptocurrency immediately.
- **Limit Orders:** Orders where the user sets a price, and the order will only execute if the purchase price is the limit price or lower, or if the sale price is the limit price or higher.

19.1.3 *Introduction*

Like securities, cryptocurrencies can be traded on platforms known as *cryptocurrency exchanges*. Binance[1] is an example of a crypto exchange.

Acquiring cryptocurrencies is akin to buying stocks on the stock exchange. Users may purchase cryptocurrencies through fiat-to-crypto exchanges. These exchanges, usually centralized, are unique as they trade crypto coins or tokens for fiat (or government-backed) currencies while other crypto exchanges trade crypto for other cryptocurrencies. IQ Option[2] is an example of a fiat-to-crypto exchange. If investors possess major cryptocurrencies such as bitcoin or ether,

---

[1] Binance (n.d.). Retrieved, February 17, 2021, from https://www.binance.com/en.

[2] iq option (n.d.). Retrieved, February 17, 2021, from https://iqoption.com/en.

they can trade these cryptocurrencies on a crypto exchange for other cryptocurrencies. Selling cryptocurrencies can also be done on exchanges with funds deposited into a connected bank account. Peer-to-peer trades or direct trades are other methods to trade cryptocurrencies.

Tether[3] is a stablecoin that is pegged to the US dollar. It has grown rapidly amidst high market volatility in 2020. It functions as the reserve currency for the crypto market. Crypto-to-crypto exchanges use Tether to price crypto assets in US dollar without having to maintain US dollar-denominated accounts, solving the difficulties in maintaining banking relationships in the crypto space.[4] Clients' funds are also held as Tether by the exchanges, reducing transaction costs until the client redeems their funds for dollars.

Decentralized Exchange (DEX) is a digital assets market that does not rely on any third party individuals or organizations to hold customer's funds; instead, the platform offers a direct peer-to-peer trading mechanism that allows the users to process their transactions on an automated system.

Margin trading, or leveraged trading, is defined as the act of borrowing to purchase an investment. This magnifies both the gains and losses on the investment.

A derivative is a contract between two parties where the value of the contract is based on an underlying asset. This allows the user to "trade" cryptocurrencies without physically owning any cryptocurrencies. Margin trading can be done using derivatives as well. Bitcoin futures allow traders to gain exposure to bitcoin without having to hold bitcoin. The Chicago Mercantile Exchange (CME) offers monthly contracts for cash settlements.[5]

---

[3]Tether (n.d.). Retrieved, February 17, 2021, from https://tether.to.

[4]Williams-Gruit, O. (2018). Everything you need to know about Tether, the cryptocurrency academics claim was used to manipulate bitcoin. Businessinsider. com. Retrieved from https://markets.businessinsider.com/currencies/news/ tether-explained-bitcoin-cryptocurrency-why-people-worried-2018-1-1014668561.

[5]Bitcoin — Electronic Platform Information Console — Confluence. (2020). Retrieved, August 26, 2020, from https://www.cmegroup.com/confluence/disp lay/EPICSANDBOX/Bitcoin.

19.1.4 *Trading-related Terms*

19.1.4.1 *General*

Long/Short: When you buy and hold a cryptocurrency, you are *long* the cryptocurrency. When you owe a cryptocurrency, you are in a *short* position. For example, if we long the ETH/BTC cryptocurrency pair, we are buying ether and selling bitcoin. We long ether and short bitcoin of an equivalent value. If we are in a long position for a cryptocurrency, our net value increases if the value of the cryptocurrency increases and decreases when the value of the cryptocurrency falls. The opposite is true if we are in a short position for the currency.

Order Types: There are multiple ways to buy and sell a cryptocurrency, market orders and limit orders being two of the more common ones. *Market orders* are orders to buy/sell the cryptocurrency immediately. The trade is guaranteed to go through, but the price is set by the market. *Limit orders* are orders where the user sets a price, and the order will only execute if the purchase price is the limit price or lower, or if the sale price is the limit price or higher.

19.1.4.2 *Market Manipulation*

Pumping occurs when a person/group purchases a massive number of assets to drive demand, which consequently increases the price of the asset. Pumpers would organize campaigns to drive enthusiasm for the cryptocurrency to instigate a purchasing frenzy. The increasing demand gives rise to the price of the cryptocurrency, which then induces more demand for the cryptocurrency. However, such manipulated increases do not stem from changes in the fundamental value of the currency and are usually followed by a dumping strategy.

After the cryptocurrency's price has reached a certain price target, the pumpers then begin a coordinated effort in the reverse direction to liquidate the cryptocurrency that they purchased while prices were lower. This set of actions is termed as *pump and dump* and often results in the pumpers earning sizeable profits while common investors may suffer from losses due to the fake price surge.

Shilling refers to people who mislead individuals into believing that a cryptocurrency is valuable and taking advantage of the rising cryptocurrency's price for his/her advantage.

19.1.4.3 *Investor Psychology*

Fear, uncertainty, and doubt (FUD) usually appear in the short-ened form in the cryptocurrency community to refer to the negative views or perceptions toward a certain event or cryptocurrency in the market.

Fear of missing out (FOMO) occurs when investors make buy-ing decisions because they feel the fear of not being a part of good cryptocurrency trends; this is the mindset that pumpers use to uneth-ically drive cryptocurrency demand further.

On the other hand, *joy of missing out* (JOMO) is when the price starts to fall (due to dumping or other reasons), and the investors that did not choose to invest in the cryptocurrency express the joy that they did not follow invest when the price was higher.

Hold on for dear life (HODL) or a HODLer describes an investor that holds a long-term investment in the cryptocurrency intending to buy and hold it. Common terms of HODLs include hodler, baghodler, ex-hodler, or landholder.

Rekt is a slang term derived from the intentional misspelling of "wrecked". In the context of cryptocurrency, rekt refers to a scenario where an investor suffered from severe or total financial loss.

19.1.4.4 *Other Terms*

Whales are used to describe people or organizations that hold sub-stantially large amounts of digital currency, stock, or other resources. As their purchasing power is magnitudes larger than a typical user, any transaction that they make would cause significant impacts to the cryptocurrency network.

Mooning is a scenario where the price of a specific cryptocurrency soars and drives the price of that cryptocurrency to new heights.

A *dip* refers to a time when the cryptocurrency has dropped in price. The phrase "Buy the dips" is a common investment strategy that advises investors to buy a cryptocurrency after it had experi-enced a dip, with the idea that its price would increase from the dip.

Airdrop is a process where a specific or existing market distributes its tokens while adhering to certain rules. When a new cryptocur-rency is created, airdrop would be a way to reach a user group.

Emission refers to the rate at which new cryptocurrency tokens are being created and used.

Faucet refers to a cryptocurrency reward system that rewards new users with cryptocurrencies with the completion of certain tasks via a website or an app. It primarily exists to raise awareness of cryptocurrencies.[6]

19.1.5   Market-related Terms

Pairs refer to the two assets we are trading. When buying or selling assets, we are always trading something of value for another of a (presumably) similar value. When buying stocks, we tend to pay in fiat currency. In that case, the pair would be a Stock/Fiat Currency. For example, ETH/BTC refers to the exchange rate of 1 ether to 1 bitcoin, also known as the *spot price*, which approximately equals 0.030639 ETH to 1 BTC today as shown in Figure 19.1.

24h Change refers to the percentage change in the exchange rate between the spot pair in the last 24 hours. In the example of ETH/BTC, Ether gained 5.43% against bitcoin over the last 24 hours.

Pair ⇅	Last Price ⇅	24h Change ⇅	24h High ⇅	24h Low ⇅	Market Cap ⇅	24h Volume ⇅
★ ETH / BTC 10x	0.030639 / $343.05	+5.43%	0.030995	0.028900	$38,414.79M	17,810.39
★ LINK / BTC 5x	0.00068370 / $7.66	+2.94%	0.00071101	0.00065417	$2,681.00M	3,606.55
★ ERD / BTC	0.00000217 / $0.024296	-13.20%	0.00000255	0.00000203	$258.25M	3,461.51
★ VET / BTC 5x	0.00000158 / $0.017690	+8.22%	0.00000166	0.00000144	$980.99M	2,253.02
★ BNB / BTC 10x	0.0018384 / $20.58	+2.22%	0.0018800	0.0017747	$3,141.86M	1,984.85
★ LTC / BTC 10x	0.005104 / $57.15	+1.63%	0.005320	0.004988	$3,723.07M	1,842.79

Figure 19.1:   Example of spot prices on Binance.[7]

---

[6]Crypto Glossary | CoinMarketCap (2020). Retrieved, August 26, 2020, from h ttps://coinmarketcap.com/glossary/.

[7]https://www.binance.com/en.

24hr High/Lows refer to the highest/lowest exchange rate in the last 24hrs between the pair. In the example of ETH/BTC, the 24hr High/Low is 0.030995 and 0.028900, respectively.

Market Capitalization (or Market Cap) is the total value of a cryptocurrency in the market. It is calculated by the current price of each cryptocurrency (usually in USD) multiplied by the circulating supply of the currency. For example, Ethereum's market cap is 38.415bn USD.

24h Volume refers to the amount of a cryptocurrency or cryptocurrency pair that were transacted in the last 24 hours. In the example of ETH/BTC, the 24hr Volume is 17,810 ethers bought or sold in exchange for bitcoin in the last 24 hours.

An *alt market* is a crypto exchange that a user can use to purchase cryptocurrencies that are not bitcoin.

Circulating supply refers to the total amount of cryptocurrency that are being held by all market participants, which could either be bought, sold, or transacted in the market. Cryptocurrencies that are retained and not transacted do not fall within the supply scope.

Maximum supply (or *Total supply*) is the maximum amount of a particular cryptocurrency that can be in circulation. For example, the maximum supply of bitcoin is about 21 million coins. The maximum supply varies between cryptocurrencies, based on their source code, or no maximum supply such as EOS.

When looking at candle charts for cryptocurrency prices, a *dildo* describes a sharp increase/decrease in price in a short period of time (usually a day). This usually represents a signal to purchase or sell the cryptocurrency. For example, Figure 19.2 shows significant increases and drops in the price.

19.2 Investment in Cryptocurrency Projects

This section explains various fundraising methods and the various differences each method has from a regulatory point of view.

19.2.1 *Learning Objectives*

- Illustrate various methods of fundraising.

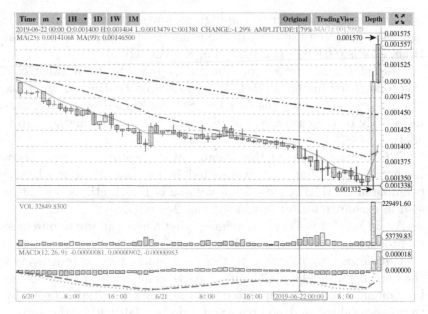

Figure 19.2:   A dildo in Aragon's (ANT) price in July 2020.

- Understand how a cryptocurrency project or start-up is started and how it raises funds.

19.2.2   *Main Takeaways*

19.2.2.1   *Main Points*

- A project should have a white paper and a roadmap before fundraising. Before an ICO, a project is not required to have a proof-of-concept.
- There are many different ways to raise funds, depending on which stage and vertical the company is positioned.
- ICOs are a cheaper, faster way for cryptocurrency projects or startups to raise funds, allowing them access to a wider pool of investors.
- ICOs are highly risky and there are currently almost no regulations in place, despite many ICO scams.
- STOs are regulation-friendly and have high potential but have not been as successful as ICOs due to certain challenges.

19.2.2.2   *Main Terms*

- **White Paper:** An informational document detailing the commercial, technological, and financial details of a new coin or project. It is typically released before an ICO and may be presented as economic, technical, or other forms of white paper.
- **Crowdfunding:** A popular mode of fundraising to attract general public funding due to its low barrier and cost of investments, diversity of projects, and a general emphasis on creativity.
- **Whitelist**: A list of all investors and stakeholders who complied with their know your customer (KYC) process.
- **Initial Crypto-Token Offering (ICO):** A swap of newly created tokens with liquid cryptocurrencies that enable blockchain startups to execute their experimental community projects.
- **Security Token Offering (STO):** A STO is similar to an ICO where an investor exchanges existing cryptocurrencies for tokens, but the security token represents an investment contract into an underlying investment asset, such as stocks, bonds, funds, and real estate investment trusts (REIT).

19.2.3   *Cryptocurrency Projects*

When investing in cryptocurrencies, we need to know the fundamental concepts in cryptocurrency projects and blockchain startups. This section explains various key terms that occur at the beginning of a cryptocurrency project.

Since cryptocurrency projects are typically technology-based, the *startup founding team* often possesses technical knowledge and skills. A founding team is not the same as founders and co-founders as founding team members are early employees. Founders are early owners. It usually consists of less than five people and its members should have aligned goals but complementary skills and knowledge. Not all members of the founding team are necessarily co-founders, and they may own very little equity as compared to the initial co-founders (Eurie, 2020).

After having an idea, the project should have a well-planned roadmap. A *roadmap* is a high-level overview of the project's major elements, such as its objectives, milestones, deliverables, strategies, and resources. This allows the team to be reminded of the overarching goal and keep track of their progress and allow investors to gauge which stage of funding the startup is positioned.

Before launching a new currency or a project, a white paper would often be released by the company. The paper consists of commercial, technological, and financial details of a new coin or project. There may be several white papers including ones focusing on economic, technical, and other aspects of the project. A white paper is highly critical in a project's fundraising process and can determine its success or failure in fundraising. Due to information asymmetry, investors rely on this paper to gain insights about the project and the team to determine whether this project is indeed worth investing in.

A *yellow paper* is more technical than a *white paper* and can be released over a white paper. It contains scientific details of the technology concisely and is typically found in startups employing deeper forms of technologies. There also exists a *beige paper*, which is simpler and more readable than yellow papers. Ethereum is one start-up that launched its ICO with all three papers.[8]

A project may also go through a *proof-of-concept* (PoC) stage, which is the first stage of idea validation. A PoC often involves a small-scale visualization exercise to test the design idea or assumption and demonstrate the feasibility of the concept (Rodela, 2020). Investors can rely on a PoC to better understand a company's offerings and gain insights into the project's feasibility. A PoC may often be confused with a *prototype*. While a PoC seeks to demonstrate that a product or feature can be developed, a prototype shows how it will be developed (Jain, 2018).

19.2.4   *Life Cycle of Token Management*

Token management's life cycle can be divided into three stages: token issue, token trading, and token burn.

Token issue: When cryptocurrency projects or startups look to raise funds, they issue tokens specific to their network in an Initial Coin Offering (ICO), in exchange for existing major cryptocurrencies such as bitcoin and ether to fund their project.

---

[8]Differences Between a White Paper, Yellow Paper, and Beige Paper. (2018). Retrieved, August 1, 2020, from https://medium.com/@hello_38248/differences-between-a-white-paper-yellow-paper-and-beige-paper-ad173f982237.

Token trading: The activity of buying and selling cryptocurrency on their respective exchanges for other assets.

Token burn: This refers to the permanent removal of existing cryptocurrency coins in circulation (Kuznetsov, 2019). It is generally used by smaller tokens for deflationary purposes, in an attempt to maintain the cryptocurrency's stable value so that current traders are incentivized to continue holding their coins. This is similar to a share buyback, where public companies purchase their stock from the market to drive up their valuation with more stocks while limiting the number of their shares available to the public. In the context of cryptocurrency, a token burn is done by sending the tokens' signature to a public wallet known as the "eater address", whose private keys cannot be obtained by any user. Burnt tokens are recorded on the ledger for all nodes to confirm that they are indeed destroyed.

19.2.5 *Fundraising Methods*

We will first introduce traditional fundraising methods by start-ups or larger companies. These include angel investments, crowdfunding, venture capital, private equity, and initial public offering (IPO).

Angel investments are typically made to well-performing projects that have demonstrated more than just their PoC. These "angels" are typically successful high net worth individuals who had similar experiences building successful companies and are now seeking an opportunity to invest back into the technology ecosystem in exchange for equity in the company. The fundraising process from angels is typically less formal than venture capitalists, as angels have significantly fewer resources to conduct comprehensive due diligence and because angels would assume that investments at this juncture would usually have less to show.

Crowdfunding is a form of early-stage investing and has provided both startups and investors with greater access to funds and investment opportunities. Crowdfunding is a relatively popular form of fundraising due to its low costs of investments, diversity of projects, and a general emphasis on supporting creative projects. In the context of cryptocurrency, projects can raise capital through an *Initial Crypto-Token Offering* (ICO), a type of crowdfunding method that would be described in later sections.

Venture Capital (VC) investments typically follow angel investments when a project/startup had demonstrated its ability to
scale. A VC fund raises capital from their *Limited Partners* (LP),
which may come in the form of pension funds, endowment funds,
large corporations, or even from angels. It is the partners' fidu-
ciary duty in a VC fund to manage returns for their LPs through
careful due diligence in the investments they make. VCs are different from angels due to their ability to write larger cheques
but rather the network and expertise that a VC team could
bring to a startup. Therefore, raising capital from VC funds
marks a level of success for startups while providing the startup with an advantage to leverage on a VC fund's expertise to
scale.

Private Equity (PE) investments typically follow VC investments and are generally the final stage of investment before
the startup goes public. Due to their generally larger sizes, PE
funds can write significantly larger cheques than VC funds and
may seek to acquire the firms they are investing in. Like VC
funds, startups raising PE capital could leverage the influence
and network of PE funds to scale their company at a faster
rate.

In the stock market, an *Initial Public Offering* (IPO) refers to a
process where a company publicly lists its shares for sale on the stock
exchange for this first time (thus going public). The purpose of such
an endeavor is to raise capital in exchange for ownership (in the form
of shares) in the firm.

A *Public Offering* is similar to an ICO in the context of cryptocurrency in a sense. The ICO stage will usually last for several days and
is divided into several funding rounds with different discounts given.
The hard cap and soft cap will then be set by the scale of financing
at a particular time.

19.2.5.1   *Initial Crypto-token Offering (ICO)*

Initial Crypto-Token Offering (ICO), also known as a *token sale* or
an *initial coin offering*, is a swap of newly created tokens with liquid cryptocurrencies that enable blockchain startups to execute their
experimental community projects. Thus, strictly speaking, token
sales are not fundraising but access to funding. It is an innovative

way to swap tokens to have access to funding and indirectly fiat currencies. Investors are not buying equity but are swapping their cryptocurrencies for tokens that are to be created by the software (Lee and Low, 2018). An emerging trend is a *pre-sale*, where investors can fund a cryptocurrency project before the official ICO. In most cases, early investors are given discounts or bonuses.[9]

ICOs started with the advent of Mastercoin[10] and were made popular after the launch of the Ethereum Network in 2015.[11]

19.2.5.1.1 Launch of ICO

ICO is perhaps one of the most successful use cases for blockchain. It allows access to international funding and immediate liquidity after investment. It has become an alternative form of crowdfunding, raising cryptocurrency instead of fiat currencies for legal entitles not limited to blockchain projects. Technology startups can have greater access to funds at a lower cost within a short time. A few projects even managed to raise hundreds of millions in a matter of a few days. Any project can launch an ICO, and any investor all around the world can invest in one.

After releasing the white paper and an official website, the cryptocurrency startup announces their project online and publishes all related information on platforms such as online forums *Reddit*, *Bitcointalk*, chat groups on Telegram, and their website. Related information published includes the ICO schedule (i.e., number, dates, and duration of the phases), the price and supply of the tokens, and the amount of funds to be raised. Information on where and how to purchase the crypto-tokens will also be published on the website or online articles/blogs.

To carry out the token swap, investors must purchase designated cryptocurrencies, usually bitcoin or ether. Investors then send the coins to the wallet address, and the ICO facilitator will email the

---

[9]What is Pre-Sale? (2020). Retrieved, August 3, 2020, from https://decryption ary.com/dictionary/pre-sale/.

[10]CoinDesk (n.d.). How does master coin work? Retrieved, February 1, 2021, from https://www.coindesk.com/crypto/mastercoin.

[11]Song, J. (2018, May 01). Why hard forks are altcoins. Retrieved from https://jimmysong.medium.com/why-hard-forks-are-altcoins-e0d3836c900d.

proof of transaction to the ICO email account. In most projects, investors will receive the tokens after the ICO ends and store them in the wallet. While some tokens will be listed on exchanges, there are many unsuccessful cases that investors will not be able to sell their tokens held and suffer from the loss. So, investors should be rational and conduct sufficient research and analysis before making any investment decisions into ICOs.

In the cryptocurrency market, there may also be caps on the amount of cryptocurrency that investors can obtain in an ICO. There are three types of caps:

- *Hard cap*: The maximum amount of tokens or funds that investors can invest in an ICO, commonly a strict upper limit of how many tokens the creators wish to sell. Most ICOs do not have hard caps.
- *Soft cap*: The minimum amount of cryptocurrencies that an investor can obtain from an ICO. The ICO is open for transactions to take place for some time. Should the ICO fail to reach the soft cap amount, funds raised would be returned to the investors.
- *Hidden cap*: An unknown limit on the amount of cryptocurrency that can be obtained from investors during the ICO. Hidden cap scenarios and limits can be implemented by the development team to prevent wealthy investors from investing too heavily, thereby allowing smaller investors to invest in a new cryptocurrency.

Cryptocurrency projects may offer a whitelisting phase for investors willing to participate in their ICO. A *whitelist* is a list consisting of all investors who complied with its know-your-customer (KYC) processes. These may be individuals, institutions, computer programs, or even cryptocurrency addresses.[12]

Return on investment (ROI) is a ratio reflecting the profitability of certain investment and is calculated by dividing the net profit by the net cost. Since 2015, bitcoin had a nearly 3,500% ROI, 70 times that of traditional stock markets.[13]

---

[12]Whitelist — Definition | Binance Academy. (2020). Retrieved, August 3, 2020, from https://academy.binance.com/glossary/whitelist.

[13]Bitcoin's ROI Since 2015 Outperforms Five Major Indices by 70X. (2020). Retrieved, August, 3 2020, from https://cointelegraph.com/news/bitcoins-roi-since-2015-outperforms-five-major-indices-by-70x#:~:text=Bitcoin%20had%20a%20nearly%203%2C500,of%20five%20traditional%20stock%20mark

19.2.5.1.2   ICO Regulation

ICOs are mostly unregulated. Thus, while ICOs have given blockchain companies access to much-needed capital, scam ICOs have become unfortunately common. In 2018, about 81% of all ICOs were scams.[14] The ease of access for startups and investors has allowed scammers to take advantage of unknowledgeable investors.

Given the speculative environment in China, large herds of amateur investors have been attracted to ICOs. The People's Bank of China (PBoC) has classified ICO as illegal public financing on September 4, 2017, citing the reason that it violates certain existing rules.[15] Though there are currently no regulations specifically for ICOs, financial centers such as the United States, England, Canada, Australia, Hong Kong, Dubai, and Singapore are issuing warnings and advisory notes to the public to warn them that some, if not all, of these ICOs, will need to conform to the existing security regulation. Switzerland and Singapore are among the few countries that have a relatively friendly environment toward the cryptocurrency market. It is important to strike a balance between managing the risks as well as encouraging innovation.

19.2.6   *Comparison: ICO, IPO, and VC*

Table 19.1 summarizes the key differences between an ICO, IPO, and VC.

19.2.7   *Security Token Offering (STO)*

A *Security Token Offering* (STO) is similar to an ICO where an investor exchanges existing cryptocurrency for tokens, but with different token characteristics. The security token represents an investment contract into an underlying investment asset, such as stocks, bonds, funds, and REIT. A security token represents ownership of these investment assets and is recorded on a blockchain.

---

ets.&text=Bitcoin%20holders'%20ROI%20is%20calculated,crypto%20to%20its%20current%20value.

[14]The Five Biggest ICO Scams. (2020). Retrieved from https://medium.com/@tozex/the-five-biggest-ico-scams-54967ec92b87.

[15]https://techcrunch.com/2017/09/04/chinas-central-bank-has-banned-icos/.

Table 19.1:   Key differences between funding via ICO, IPO[a] and VC.[b]

	ICO	IPO	VC
Investor's rights	New tokens that entitle them to certain rights, but not shares of the legal entity. Tokens may not be deemed as securities	Equity ownership and voting rights in the form of a security	Get a seat on the board, allowing them to monitor and advise the startup
Regulations	No requirements or standards on disclosure before initiation of ICO, mostly unregulated	Strict regulatory filings and disclosures required	VCs have proper due diligence procedures to ensure that they are investing in legitimate companies
Stage of project/ company	Early-stage, able to raise funds with just an idea	Mature, required to fulfill certain requirements relating to scale and profitability	Typically demand a PoC and a knowledgeable and experienced management team
Type of investors	Open to all types of investors globally	Mostly institutional investors	Professional investors who usually come from the same country or region

(Continued)

Table 19.1:   (*Continued*)

	ICO	IPO	VC
Type of investment	The liquid investment provides an easy exit strategy for investors	Liquid investment as shares can be traded on the secondary market	Illiquid investment and more long-term commitment
Limit on funds raised	Usually comes with an upper limit, expressed as a maximal number of tokens to be sold	No upper limit on funds raised	Depends on the VC's funds and the investment committee's judgement

Sources: [a]ICO vs IPO: Major differences |Hacker Noon. (2018). Retrieved, August 26, 2020, from https://hackernoon.com/ico-vs-ipo-major-differences-bd23890cb83b.
[b]Mukhopadhyay, A. (2018). ICO vs VC: The experience matters. Retrieved from https://101blockchains.com/ico-vs-vc-comparison/.

Since STOs are backed by tangible assets, it is easy to assess their fair value relative to these underlying assets. The greatest advantage of an STO is its compliance with securities laws. They are also cheaper and faster than IPOs due to smart contracts and blockchain removing middlemen like underwriters and the need for paperwork. In the secondary market, security tokens can be traded round the clock, as compared to the fixed trading hours for the secondary market after an IPO. Thus, STOs can be seen as a hybrid between ICOs and IPOs, combining the efficiency of ICOs and the regulatory compliance of IPOs.[16]

---

[16]Pauw, C. (2019). What is an STO, explained. Retrieved from https://cointelegraph.com/explained/what-is-an-sto-explained.

Table 19.2: Differences between an ICO, IPO, and STO.[a,b]

	ICO	IPO	STO
Underlying assets	No collateral	Backed by tangible assets	Backed by tangible assets
Regulations	No compliance require-ments	Strict regulatory require-ments	Compliant with securities laws
Risk	High risk	Low risk	Moderate risk
Investor pool	Open to all investors across all geographies	Investors are geographi-cally restricted	Accredited investors
Investor rights	Access to network, platform, or service provided by the project	Equity ownership	Equity ownership and access to network, platform, or service provided by the project
Stage of project/ company	Projects or start-ups with an idea	Established company with profitability	Companies issuing the underlying assets

Sources: [a]The Difference Between ICO and STO. (2020). Retrieved, August 26, 2020, from https://cryptonews.com/guides/the-difference-between-ico-and-sto.htm.
[b]ICO vs STO: All You Need to Know About the New Fundrais-ing Method in the Crypto World. |Hacker Noon. (2019). Retrieved, August 26, 2020, from https://hackernoon.com/ico-vs-sto-all-you-need-to-know-about-the-new-fundraising-method-in-the-crypto-world-54a1 a43a08d6.

Table 19.2 summarizes the key differences between ICOs, IPOs, and STOs.

19.2.7.1 *Challenges of STOs*

STOs possess huge potential but currently face multiple challenges that inhibit their mainstream success as compared to ICOs. First, although STOs allow global investors with easy access, the current regulatory environment in each country is unclear and limited to periodic guidance. This hinders the long-term viability of STOs. Second, regulations limiting the investor pool and capital raised may deter start-ups from launching an STO. Third, although there are active developments in STOs, this type of offering is still in its infancy and lacks comprehensive standards that hinder an effective execution. Fourth, investors need to be convinced that besides inherent security risks, the additional cost (associated with legal, custody, and regulatory) and technology risk will not be prohibitive. However, intermediaries such as crypto banks and custody, as well as regulatory clarity, are lowering the barrier to STOs.

19.2.8   *Other Fundraising Methods*

Here are a few less commonly known fundraisings events:

Initial Exchange Offering (IEO): As its name suggests, IEOs are fundraising events administered by exchange, in contrast with an ICO where the project team administers the fundraising.[17] It is easier and possibly less risky for users to participate in an IEO as compared to an ICO.

Initial Bounty Offering (IBO): An IBO refers to openly issuing a new cryptocurrency for a set period in exchange for services rather than actual money. This type of offering potentially prevents large buyers from manipulating the market.

Initial Miner Offering (IMO): Unlike ICOs and IPOs where the intention is to directly fund the company in question, IMOs are fundraisings for mining organizations. As Proof-of-Work consensus rewards the fastest miner, mining operations have become more institutionalized as only corporates have the capital to purchase leading technologies and bargain for efficient electrical prices. Such companies leverage on IMOs to fundraise their initial expenditures.

Initial Fork Offering (IFO): In this case, a hard fork occurs on an existing cryptocurrency with some pre-mined "new" cryptocurrencies being sold to investors (Lambert *et al.*, 2020; OECD, 2019).

---

[17]Initial Exchange Offering (IEO) — Definition |Binance Academy. (2020). Retrieved, August 3, 2020, from https://academy.binance.com/glossary/initial-exchange-offering.

Bibliography

Eurie, K. (2020). Founder vs. founding team member. Stanford eCorner. Retrieved from https://ecorner.stanford.edu/in-brief/founder-vs-foun ding-team-member/.

Fang, F., Ventre, C., Basios, M., Kong, H. *et al.* (2020). Cryptocurrency trading: a comprehensive survey. arXiv preprint arXiv:2003.11352.

Jain, M. (2018). What is the difference between proof of concept and prototype? Retrieved from https://www.entrepreneur.com/article/307454.

Kuznetsov, N. (2019). Token burning explained. Retrieved from https://c ointelegraph.com/explained/token-burning-explained.

Kyriazis, N. A. (2019). A survey on efficiency and profitable trading opportunities in cryptocurrency markets. *J. Risk Financ. Manag.*, 12, 67.

Lambert, T., Liebau, D., and Roosenboom, P. (2020). Security token offerings. Working paper.

Lee, J. Y. (2019). A decentralized token economy: How blockchain and cryptocurrency can revolutionize business. *Business Horizons*, 62(6), 773–784.

Lee, D. K. C. and Low, S. (2018). *Inclusive Fintech: Blockchain, Cryptocurrency, and ICO*. World Scientific.

Mukhopadhyay, U., Skjellum, A., Hambolu, O., Oakley, J. *et al.* (2016). A brief survey of cryptocurrency systems. In *2016 14th Ann. Conf. Privacy, Security and Trust (PST)*, IEEE, pp. 745–752.

OECD (2019). Initial coin offerings (ICOs) for SME financing. Retrieved from www.oecd.org/finance/initial-coin-offerings-for-sme-financing. htm.

Pauw, C. (2019). What is an STO explained? Cointelegraph. Retrieved from https://cointelegraph.com/explained/what-is-an-sto-explained.

Rodela, J. (2020). How to create a proof of concept in 2020. Retrieved from https://www.fool.com/the-blueprint/proof-of-concept/.

Sharma, S., Krishma, N., and Raina, E. (2017). Survey paper on cryptocurrency. *Int. J. Sci. Res. Computer Science, Engineering and Information Technology*, 2(3), 307–310.

19.3   Sample Questions

Please select the most appropriate response.

Question 1

A new cryptocurrency, CryptoX, has the following properties:

Crypto X (as at January 31, 2020)

Circulating Supply	10,000
Maximum Supply	500,000
Current Price	$3
24hr High/Low	$5/$2
24hr Volume	500

Calculate the Market Cap of CryptoX:

(a) $1,500,000
(b) $30,000
(c) $1,000,000

Question 2

Which of the following orders would have been executed if they were placed at the beginning of the trading period January 31, 2020:

Order	Type	Target Price
1 Buy	Market	n/a
2 Buy	Limit	1
3 Sell	Market	n/a
4 Sell	Limit	5

(a) 1,3
(b) 1,3,4
(c) 1,2,3,4

Question 3

Which of the following are true about pumping and dumping?

(a) Pump and Dump is illegal for cryptocurrency and therefore does not happen often.
(b) Pump and Dump is not officially illegal for cryptocurrency and therefore investors need to be wary of such schemes.

(c) Pump and Dump is the act of hyping up the prices of a cryptocurrency so that the prices of the asset keep rising.

Question 4

What is one of the reasons that a proof-of-concept is needed?

(a) To allow the project team to test how much resources are needed to build their product.
(b) To increase their credibility so that investors would invest when the project team launches an ICO.
(c) To allow the project team to test the methodology of building their product.

Question 5

What type of investor should invest in ICOs?

(a) Any investor around the world who have a cryptocurrency wallet with existing cryptocurrencies that can be swapped with the token issued in the ICO.
(b) Investors who are familiar with blockchain and cryptocurrency, and are careful in assessing whether the team is credible.
(c) Investors who want to make quick profits.

Question 6

Despite STOs combining the benefits of both ICOs and IPOs, what is one of the reasons STOs they not as popular as ICOs?

(a) STOs have to comply with securities laws.
(b) STOs have not found product market fit.
(c) STOs are more costly.

Solutions

Question 1

Solution: Option **b** is correct.

Market Cap is calculated by multiplying circulating supply with current price.

Question 2

Solution: Option **b** is correct.

A market order always executes at the prevailing market price. Order 2 does not occur as the Limit buy will only execute if the price of the cryptocurrency is equal or LESS than the target price.

Question 3

Solution: Option **b** is correct.

Pump and Dump is not officially illegal as cryptocurrency is still an emerging asset and has a lack of official regulation yet to protect investors. C is wrong as it only describes the pumping scheme and not the dumping scheme where the scammers liquidate their shares of the overpriced assets.

Question 4

Solution: Option **a** is correct.

A proof-of-concept is used to test the feasibility of a project—whether it can work, how much resources it will consume, and whether it is worth it to continue with the idea.

Question 5

Solution: Option **b** is correct.

Although key benefits of ICOs are that there are not geographically restricted for investors and possess a high potential for capital appreciation, investing in ICOs is highly risky and a vast majority of ICOs turn out to be fraudulent. Only knowledgeable investors who can determine the authenticity of an ICO project should invest.

Question 6

Solution: Option **b** is correct.

Although the concept of STOs sounds promising and innovative, there are misaligned incentives between key market participants. The regulatory compliance requirements are not an obstacle for the adoption of STOs, as all companies issuing securities have to comply with those laws. The type of companies raising funds for an ICO and an STO is different.

Global Fintech Institute - World Scientific Series on Fintech

(Continued from page ii)

Applications and Trends in Fintech II: Cloud Computing, Compliance, and
Global Fintech Trends
> David Lee Kuo Chuen (Global Fintech Institute, Singapore & Singapore
> University of Social Sciences, Singapore), Joseph Lim, Phoon Kok Fai and
> Wang Yu (Singapore University of Social Sciences, Singapore)